"中国森林生物多样性监测网络"丛书　马克平 主编

鼎湖山南亚热带森林
——树种及其分布格局

Dinghushan Lower Subtropical Forest Dynamics Plot: Tree Species and Their Distribution Patterns

曹洪麟　吴林芳　王志高　黄忠良
李　林　魏识广　练琚愉　叶万辉　著

中国林业出版社
China Forestry Publishing House

图书在版编目（CIP）数据

鼎湖山南亚热带森林——树种及其分布格局 / 曹洪麟等著. —北京：中国林业出版社，2013.4
（中国森林生物多样性监测网络丛书）

ISBN 978-7-5038-7049-1

I. ①鼎… II. ①曹… III. ①亚热带－森林－树种－研究－广东省 IV. ①S717.265

中国版本图书馆CIP数据核字(2013)第103345号

内容简介

本书介绍了广东省鼎湖山南亚热带常绿阔叶林常见木本植物195种，每种植物除了文字描述外，还配有精美的照片，展示植物的植株、花序、果实或幼苗等，方便识别。同时附有每种木本植物在20hm^{2}长期定位研究样地种群分布图以及种群的个体数量和径级结构。对于该样地的地形、土壤、植被等皆有介绍。本书以资料翔实、图片精美为特点，是南亚热带常绿阔叶林研究不可多得的参考书，也可以作为植物爱好者了解南亚热带森林、认识森林植物的野外指导手册。

中国林业出版社·环境园林图书出版中心

策划、责任编辑：于界芬 盛春玲
电话：83229512 传真：83227584

出 版：中国林业出版社（100009 北京西城区德内大街刘海胡同7号）
网 址：http://lycb.forestry.gov.cn/
发 行：中国林业出版社
印 刷：北京卡乐富印刷有限公司
版 次：2013年8月第1版
印 次：2013年8月第1次
开 本：889mm × 1194mm 1 / 16
印 张：13.75
字 数：352千字
定 价：138.00元

序 言 1

在过去的几十年时间里，中国科学院和林业、农业等相关部门陆续建立了数百个生态系统定位研究站。其中，中国科学院组建的中国生态系统研究网络（CERN）拥有分布于全国包括农田、森林、草地、湿地、荒漠等生态系统类型的36个生态站。国家林业局建立的中国森林生态系统研究网络（CFERN）由29个生态站组成，基本覆盖了我国典型的地带性森林生态系统类型和最主要的次生林、人工林类型。

随着研究的发展，特别是近年来人们对生物多样性和全球变化研究的关注，国际上正在推动生态系统综合研究网络平台的建立。在全球水平上，全球生物多样性综合观测网络（GEO-BON）是一个有代表性的研究网络。它试图把全球与生态系统和生物多样性长期定位研究相关的网络整合起来，通过综合研究，探讨生态系统与生物多样性维持与变化机制以及系统之间的相互作用机理，为生态系统可持续管理与生物多样性的保护提供科学依据和管理模式。

近年来，中国科学院生物多样性委员会组织建立了中国森林生物多样性监测网络（Chinese Forest Biodiversity Monitoring Network，以下简称 CForBio）。中国是生物多样性特别丰富的少数国家之一，也是唯一一个具有从北部寒温带到南部热带完整气候带谱的国家。CForBio对于揭示中国森林生物多样性形成和维持机制，以及森林生物多样性对全球变化的响应，科学利用和有效保护中国森林生物多样性资源具有重要意义。

截止到2012年，中国森林生物多样性监测网络包括大型监测样地12个。12个建成样地具体如下：黑龙江小兴安岭丰林阔叶红松林30hm^{2}样地、黑龙江小兴安岭凉水典型阔叶红松林9hm^{2}样地和谷底云冷杉林9hm^{2}样地、吉林长白山阔叶红松林25hm^{2}样地、北京东灵山暖温带落叶阔叶林20hm^{2}样地、河南宝天曼暖温带落叶阔叶林25hm^{2}样地、湖南八大公山中亚热带山地常绿落叶阔叶混交林25hm^{2}样地、浙江天童亚热带常绿阔叶林20hm^{2}样地、浙江古田山亚热带常绿阔叶林24hm^{2}样地、广东鼎湖山南亚热带常绿阔叶林20hm^{2}样地、广西弄岗喀斯特季节性雨林15hm^{2}样地、云南西双版纳热带雨林20hm^{2}样地。目前，正在与黑龙江省科学院自然与生态研究所合作建立中国寒温带大兴安岭兴安落叶松林25hm^{2}样地，并与中科院武汉植物园达成协议，拟在2013年建立温带－亚热带过渡区秦岭落叶阔叶林25hm^{2}样地。此外，还有若干5hm^{2}的样地，如吉林长白山白桦林、北京东灵山辽东栎林、浙江百山祖常绿阔叶林和四川都江堰常绿落叶阔叶混交林样地。CForBio是继美国史密森研究院热带研究所建立的热带森林生物多样性监测网络（CTFS）之后又一大型区域生物多样性监测网络。由于CForBio横跨多个纬度梯度，对揭示生物多样性科学的基本规律有特殊意义，在国际生物多样性监测网络中具有重要地位。

目前，CForBio已经有很好的研究进展，各样地研究成果陆续在国际著名生态学刊物如*Ecology, Journal of Ecology, Oikos*等上发表，受到国内外同行的高度评价。但这些文章都是关于某一具体问题的研究总结，还无法让国内外同行全面了解CForBio各个样地整体情况。因此，出版这套以中英文形式介绍各大样地基本情况的“中国森林生物多样性监测网络”丛书是非常必要的。感谢马克平研究员组织相关专家编写这套丛书。我相信该丛书不仅是国内外同行深入了解CForBio各样地的参考书，同时也将为我国森林生物多样性监测和森林生态系统联网研究奠定重要的基础。

Foreword 1

In the past few decades, hundreds of Ecosystem Research Stations have been set up by the Chinese Academy of Sciences, State Forestry Administration, Ministry of Agriculture and other relative departments. Among them, 36 ecological research stations were established by Chinese Ecosystem Research Network (CERN), supported by the Chinese Academy of Sciences. The 36 research stations are scattered over the country representing diverse ecosystems, including farmland, forest, grassland, wetland, desert and others. Moreover, the Chinese National Ecological Research Network (CFERN), supported by the State Forestry Administration, consists of 29 research stations, covering typical zonal forest ecosystems and main secondary forests and plantations in China.

With the development of research, especially the growing concern over researches on biodiversity and global change in recent years, the establishment of ecosystem research network have been promoted under international supports. So the Group on Earth Observations Biodiversity Observation Network (GEO-BON) is representative across the world, and it attempts to integrate worldwide networks relating to long-term research on ecosystem and biodiversity. Based on the comprehensive studies, the maintenance and change mechanism of ecosystem and biodiversity and their interactions have been explored, which provide scientific basis and management mode for sustainable development of ecosystem and protection of biodiversity.

In recent years, Chinese Forest Biodiversity Monitoring Network (CForBio) has been built by Biodiversity Committee of the Chinese Academy of Sciences. China is one of the few top “mega-biodiversity countries” in the world, and it is also the only country with full climatic zone spectrum, ranging from northern cool temperate zone to southern tropical zone. Besides, CForBio is of great significance to reveal the formation and maintenance mechanism of forest biodiversity in China and their response to climate change.

Twelve permanent plots with area more than 9 hm^2 have been set up by CForBio till 2012. The 12 permanent plots are as follows: 30 hm^2 mixed broadleaved Korean pine forest plot at Fenglin in Xiaoxing'an Mountains in Heilongjiang; 9 hm^2 mixed broadleaved-Korean pine forest plot and 9 hm^2 Spruce-fir valley forest plot at Liangshui in Xiaoxing'an Mountains in Heilongjiang; 25 hm^2 deciduous broadleaved Korean pine forest plot at Changbai Mountain in Jilin; 20 hm^2 deciduous broadleaved forest plot at Dongling Mountain in Beijing; 25 hm^2 deciduous broadleaved forest plot at Baotianman in Henan; 25 hm^2 mid-subtropical mountain evergreen and deciduous broadleaved mixed forest plot at Badagong Mountain in Hunan; 20 hm^2 subtropical evergreen broadleaved forest plot at Tiantong Mountain in Zhejiang; 24 hm^2 subtropical evergreen broadleaved forest plot at Gutian Mountain in Zhejiang; 20 hm^2 lower subtropical evergreen broadleaved forest plot at Dinghu Mountain in Guangdong; 15 hm^2 karst seasonal rain forest plot at Longgang in Guangxi; 20 hm^2 tropical rain forest plot at Xishuangbanna in Yunnan. A 25 hm^2 dahurian larch forest plot in Daxinganling in Heilongjiang is now under construction in the collaboration with Institute of Natural Resources, Heilongjiang Academy of Sciences. Wuhan Botanical Garden, CAS, is going to establish a 25 hm^2 deciduous broadleaved forest plot in Qinling Mountain in Shannxi. Besides, a number of 5 hm^2 plots have also been built, including birch forest plot in Changbaishan, Jiling province, oak forest plot in Donglinshan, Beijing, evergreen broadleaved forest plot in Baishanzu, Zhejiang province and evergreen and deciduous broadleaved mixed forest plot in Dujiangyan, Sichuan province. Now, CForBio is another regional biodiversity monitoring network after the Center for Tropical Forest Science (CTFS). As being across several latitudinal gradients, CForBio is not only important to examining the fundamental mechanism of biodiversity maintenance, but also plays an important role in Global Biodiversity Monitoring Network.

Encouraging progress has been made in this area since the network built, for lots of research findings have been published in the international peer reviewed ecological journals, such as *Ecology, Journal of Ecology* and *Oikos*, etc., which brought about positive response from colleagues in the field of plant ecology. However, the published papers mostly focus on research of specific problems; scientists and public still can't understand the whole situation of each plot in details. So it is really necessary to publish this series, which introduce basic information of permanent forest plots in both Chinese and English. I am grateful to Professor Keping Ma for organizing related specialists to prepare the series. And I believe that this series would be a valuable reference book for scientists and public to further understand CForBio, and it will also lay a foundation for the forest biodiversity monitoring and forest ecosystem research in China.

Honglie Sun
The former Vice-President for the Chinese Academy of Sciences

序 言 2

森林在维持世界气候与水文循环中起着根本性的作用。森林是极为丰富多样的动物、植物与微生物的家园，而人类正是依靠这些生物获取各种产品，包括食品与药物。尽管对人类福祉如此重要，森林仍然遭受着来自土地利用与全球气候环境变化的巨大威胁。在这种不断变化的情况下，为了更好地管理全球剩余的森林，迫切需要树种在生长、死亡与更新方面的详细信息。

中国森林生物多样性监测网络（CForBio）正在中国沿着纬度与环境梯度建立大尺度森林监测样地。通过这个重要的全国行动倡议与来自中国科学院及若干其他单位的研究者的努力，CForBio开始搜集关于中国森林的结构与动态的关键信息。现在CForBio与史密森研究院及哈佛大学阿诺德树木园的热带森林监测网络（CTFS）形成了合作伙伴。CTFS是个在21个热带或温带国家拥有长期大尺度森林动态研究样地的全球性网络。CForBio与CTFS合作的目标是通过合作研究，了解森林是如何运作的，它们是如何随着时间而改变的，以及如何重建或者恢复，以确保森林提供的环境服务能可持续或者增长。森林及其提供的服务的长期可持续性有赖于我们预测森林对全球变化，包括气候与土地利用变化的响应的能力，以及我们去理解与创建适当的森林服务市场的能力。通过拥有34个森林大样地的全球网络及大量项目的训练与能力建设，CForBio与CTFS的伙伴关系是发展这些预测工具的重要基础。这种伙伴关系也将促进为全球各地的当地社区、林业管理者与政策制定者在森林的保育与管理方面发展应用性的林业项目建议，发展与示范利用乡土物种进行森林重建的方法，以及从经济学角度评估森林在减缓气候变化、生物多样性保护和流域保护上的价值的方法。

我祝贺作者们创作了这部关于样地植物的优秀丛书。本丛书为将来的森林监测提供了基准信息，是涉及森林恢复、碳存储、动植物关系、遗传多样性、气候变化、局地与区域保育等研究内容的研究者、学生与森林管理者们有价值的参考资料。

S. J. 戴维斯
主任
史密森热带研究所&哈佛大学阿诺德
树木园热带森林科学研究中心

Foreword 2

Forests play an essential role in regulating of world's climatic and hydrological cycles. They are home to a vast array of animal, plant and microorganism species on which humans depend for many products, including food and medicines. Despite the importance of forests to human welfare they are under enormous threat from changes in land-use and global climatic conditions. In order to better manage the world's remaining forests under these changing conditions detailed information on the dynamics of growth, mortality and recruitment of tree species is urgently needed.

The Chinese Forest Biodiversity Monitoring Network (CForBio) that aims to establish large-scale forest monitoring plots across latitudinal and environmental gradients in China. Through this important national initiative, researchers from the Chinese Academy of Sciences and several other research institutions in China, CForBio has begun to gather key information on the structure and dynamics of China's forests. The CForBio initiative is now partnering with the Center for Tropical Forest Science (CTFS) of the Smithsonian Research Institute and the Arnold Arboretum of Harvard University. CTFS is a global program of long-term large-scale forest dynamics plots in 21 tropical and temperate countries. The goal of the partnership between CForBio and CTFS is to work together to understand how forests work, how they are changing over time, and how they can be re-created or restored to ensure that the environmental services provided by forests are sustained or increased. The long-term sustainability of forests and the services they provide depend on our ability to predict forest responses to global changes, including changes in climate and land-use, and our ability to understand and create appropriate markets for forest services. The CForBio-CTFS partnership is ideally poised to develop these predictive tools through a global network of 34 large forest plots and an extensive program of training and capacity building. The partnership will also lead to the development of applied forestry programs that advise local communities, forest managers and policy makers around the world on conservation and management of forests, to develop and demonstrate methods of native species reforestation, and to economically value the roles that forests play in climate mitigation, biodiversity conservation, and watershed protection.

I congratulate the authors on the production of this excellent new series of stand books. In addition to providing a baseline for future forest monitoring, these books provide a valuable resource for researchers, students, and forest managers dealing with issues of forest restoration, carbon storage, plant-animal interactions, genetic diversity, climate change, and local and regional conservation issues.

Stuart Davies

Director

Center for Tropical Forest Science / SIGEO

The Smithsonian Tropical Research Institute &

The Arnold Arboretum of Harvard University

前　言

常绿阔叶林是指在亚热带湿润季风气候条件下，由常绿阔叶树组成的森林植被。分布于地球表面热带以北或以南中纬度亚热带区域，在北半球其分布位置大致在北纬22º～34º（40º）。在欧亚大陆东南部，常绿阔叶林主要分布于中国的长江流域至珠江流域一带；朝鲜半岛、日本列岛南部也有分布。此外，非洲的东南沿海和西北部大西洋中的加那利与马德拉群岛；北美洲的东南端和墨西哥；南美洲的智利、阿根廷、玻利维亚及巴西的一部分；大洋洲东岸以及新西兰的北岛等地均有常绿阔叶林分布，其中，以中国长江流域至珠江流域的常绿阔叶林区最为典型，面积也最大。

常绿阔叶林以其富饶的生物资源、丰富的物种多样性和巨大的环境效益，引起世人越来越多的关注，对它的研究已成为国际植被科学界关注的主题之一。我国对常绿阔叶林的研究，可追溯到20世纪30年代胡先骕、钱崇澍、樊庆笙等对我国南方植被的描述，但在1949年以前只有一些零星的相关论文发表。具有一定规模的研究始于20世纪50年代，当时的研究大都结合热带、亚热带生物资源考察，如1952～1955年华南垦殖勘察，1958～1961年华南热带生物资源综合考察，1957～1961年云南热带、亚热带生物资源综合考察，以及一些省区组织的野生生物资源考察（如福建省1957年的考察）等。这些调查研究积累了很多有关常绿阔叶林群落的组成结构和类型分布的丰富资料，并陆续出版了《中国植被》和各省植被专著，但此时的常绿阔叶林研究仍处于以定性调查为主的阶段。

随着一批有关数量生态学和种群生态学等著作的引入和翻译，近三十年来我国在常绿阔叶林方面做了不少定量研究工作，对常绿阔叶林的物种组成、空间结构、分布特征和生态系统功能等开展了一定的研究，积累了相当丰富的资料和经验。但是这些数据仅局限于少数几个定位研究站（如广东的鼎湖山、重庆的缙云山、浙江的天童山等），是在小面积样地（一般小于1hm^2）和短时间监测等基础上获得的，难以涵盖众多的稀少物种和不同的生态环境，较难反映群落内不同尺度上的生物多样性格局和过程，不足以认识群落水平上生物多样性的维持机制。因此，有必要在我国广大的常绿阔叶林地区的不同地带，建立相应的大面积长期监测样地，对常绿阔叶林从植物开花与结实、种子形成与扩散、幼苗更新与定居、种群建立与维持、个体生长与死亡的整个过程进行监测研究，分析影响这些过程的各种生物和非生物因素，探讨常绿阔叶林生物多样性形成与维持的机制。

中国科学院鼎湖山国家级自然保护区建立于1956年，是我国最早建立的自然保护区，保存有400多年历史的地带性森林群落，是开展南亚热带常绿阔叶林生物多样性形成与维持机制研究的理想场所。2004～2005年，在中国科学院生物多样性委员会的支持下，我们参照CTFS（Centre for Tropical Forest Science）样地建设标准，在鼎湖山庆云寺后山至三宝峰之间保存最好的南亚热带常绿阔叶林内，建立了20hm^{2}的永久监测样地，以期对南亚热带常绿阔叶林生物多样性进行长期监测与研究。按照5年复查一次的要求，我们于2010年完成了对大样地的第一次复查，并在大样地周边地区建立了5个不同群落类型1hm^{2}的附属样地，附属样地的建设标准与大样地相同。同时根据复查结果对2005年的调查数据进行了修正。

本书详尽描述了修正后的鼎湖山20hm^{2}大样地（DHS大样地）树种的分布格局、径级结构和生物学特性，容易混淆树种的野外识别特征等，为今后的深入研究提供了必需的基本信息。而精美的样地植物照片将对读者认知鼎湖山森林和植物提供更加感性的材料，以期吸引更多的年轻学子加入森林生态学的探索行列。

在地形复杂的鼎湖山建设20hm^{2}大样地，是一个前所未有的挑战。在样地建设及后来的科研活动过程中，得到了方方面面的大力支持与帮助。借此机会，对这些单位和同事为鼎湖山样地建设及研究作出的贡献表示诚挚的感谢：加拿大Alberta大学何芳良教授、东华大学孙义方教授、中国科学院植物研究所马克平研究员，不仅在建设样地过程中提出宝贵意见，并且在样地选址、调查工作中亲临现场给予指导；美国热带森林研究中心（CTFS）生态学家Richard Condit教授在数据整理上给予重要的指导和帮助；武汉江韵勘测工程集团有限公司精确与辛勤地完成了样地的测点工作；中国科学院华南植物园的黄忠良研究员、陈炳辉高工、练琚愉副研究员、黄玉佳、莫定升、孟泽、张佑昌、吴林芳、叶育石、陈银洁、蔡文波、方晓明、向传银，华南师范大学林正媚博士，浙江古田山自然保护区的方腾，中国科学院华南植物园研究生：王志高、史军辉、张池、李林、魏识广、沈浩、李静、穆宏平、韩玉洁、林国俊、刘文平、廖凌娟、李小意、马磊、沈勇、许淑君、李博文、王兰英、牛红玉、王冉、董蕾；华南师范大学、华南农业大学与广州中医药大学本科生：施悦谋、陈达丰、谢腾芳、黄婷、陈裕喜、李胜强、陈水莲、温远香、郭阿琴、庞冯连、杨国强、梁国晖、刘肇基、古敬锋、何淑兰、李新达等，在样地建设、植物调查与复查、数据录入等方面付出了许多艰辛的劳动；吴林芳、叶育石、林正媚、董安强等为本书提供了许多精美的照片；中国科学院华南植物园叶华谷研究员为本书编写提出了许多宝贵的意见。

本书收录了鼎湖山大样地木本植物195种（含变种），隶属于55科111属。本书裸子植物按郑万钧1975年系统、被子植物按哈钦松系统排列。植物的生物学特性描述主要参考《中国植物志》和*Flora of China*。

由于时间仓促，水平有限，错误疏漏在所难免，敬请各位读者不吝赐教。

叶万辉　曹洪麟

2011年12月

Preface

Evergreen broadleaved forests are the vegetations consisting of evergreen broadleaved trees in the subtropical zones with moist monsoon climate. They distribute in the north and south of the equator, and in the Northern Hemisphere between 22° and 34° (40°) N. In southeast Eurasia, they mainly occur in Yangtze River and Pearl River Basins in China, and they are also found on Korean Peninsula and in southern Japan. In addition, they also grow in the regions along the sea in southeast Africa, on the Canary and Madeira Islands in northwest Atlantic, in Mexico and southeast of North America, in Chile, Argentina, Bolivia and Brazil in South America, and in the east coast of the Oceania and the northern islands of New Zealand. Among all these, the forests in China are the most typical and cover the largest area.

With rich biotic resources, abundant biodiversity and enormous environmental benefits, the evergreen broadleaved forests have attracted the attention of the world, and they have become the major research targets in the scientific community vegetation science in the world in decades. Studies on them in our country can be traced back to the 1930s when Xiansu Hu, Chongshu Qian, Qingsheng Fan, et al., described the vegetations in South China, but there were only a small number of relevant papers published sporadically before 1949. In the 1950s, a large number of scientific research projects were initiated, and they were mainly combined with the biological resource surveys in the tropical and subtropical zones, such as the cultivation survey in South China from 1952 to 1955, the comprehensive survey of the tropical biological resources in South China from 1958 to 1961, the comprehensive survey of the tropical and subtropical biological resources in Yunnan Province from 1957 to 1961, and the surveys of wild biological resources in other provinces (i.e., the survey in Fujian province in 1957), and so on. These surveys collected much information on the structure, types, and distributions of evergreen broadleaved forests and published "Vegetation of China" and documentations on vegetations in different provinces continuously. However, these surveys and researches were descriptive, and this period may be considered a qualitative phase in the studies of such forests in China.

Following the introductions and translations of foreign references on quantitative and population ecology, many quantitative researches were conducted on evergreen broadleaved forests in the last 30 years in China focusing on the species composition, spatial structure, distribution characteristics and ecological functions. Consequently, much data and experience were obtained. However, these data are limited to the forests in several forest ecosystem research stations (i.e., Dinghushan in Guangdong, Jinyunshan in Chongqing, Tiantongshan in Zhejiang) based on small sampling areas, usually smaller than 1 hm^2 and short-time observations. It is impossible for them to contain many rare species and environmental conditions and to understand biodiversity patterns and processes at different spatial scales in a forest community. Therefore, they are insufficient to understand the mechanisms of biodiversity maintenance at community level. Thus, it is necessary to establish large-scale and long-term evergreen broadleaved forest dynamic plots in different geographical zones in China, to observe and study the whole processes from flowering to fruiting, from seed formation to dispersal, from seedling recruitment to establishment, from population establishment to maintenance, and from individual growth to death of plants, to analyze the biotic and abiotic factors influencing these processes and to explore the mechanisms of biodiversity formation and maintenance of the evergreen broadleaved forests.

The Dinghushan National Nature Reserve of the Chinese Academy of Sciences (CAS) was founded in 1956. It is the earliest one in China. Its vegetation has been protected for more than 400 years, and it is representative of its geographical zonal vegetations. Thus, the Reserve is an ideal place to study the mechanisms of biodiversity origin and maintenance of evergreen broadleaved forests in the south subtropical zones. From 2004 to 2005, following the recommendations of the 2004 Beijing Workshop and with the support of the Biodiversity Committee of CAS, we, the conservation ecology research group at the South China Botanical Garden, established the Dinghushan 20 hm^2 forest dynamic plot (DHS Plot) in the best protected south subtropical evergreen broadleaved forests between Qingyun Temple and Sanbao Mountain in the Dinghushan Reserve following the standard census protocol of CTFS (Center for Tropical Forest Sciences), with the hope of monitoring and studying the biodiversity of south subtropical evergreen broadleaved forests on a long term basis. According to the census standard of surveying once every five years, we conducted the first re-census of the DHS plot in 2010. At the same time, we established five associated 1 hm^2 plots with different types of forest communities in the adjacent areas along the edges of the DHS plot using the same standards as those used for the DHS plot. Meanwhile, we corrected the survey data of 2005 based on our first re-census.

This book describes in details the distribution patterns, DBH structures and biological characteristics of the tree species based on the corrected survey data and the field identification guides for the trees that can be easily misidentified in the DHS plot and its associated plots. It provides the necessary baseline information for further studies. It contains beautiful photos of the plants in the plots, and these provide readers with perceptual materials to know the forests and plants in the plots. All these will help to attract more and more the young and educated to join the research team in forest ecology.

It was an unprecedented challenge to establish the large 20 hm^2 forest dynamic plot in terrain-rugged Dinghushan. During the establishment of the plot and research work carried out in it later, many people and organizations gave us great support and help. Hence, we would like to take this opportunity to greatly acknowledge the followings for their admirable contributions. Professor Fangliang He in University of Alberta, Canada, Professor I-Fang Sun in Dong Hwa University and Professor Keping Ma in the Institute of Botany of CAS not only provided us with valuable ideas for establishing the plot, but also gave us help personally in the location selection of the plot and census work. Dr. Richard Condit in CTFS gave us important guidance and assistance in data management. With hard work, Jiangyun Surveying and Mapping Exploration Co. Ltd. accurately measured and located the geographical coordinates for the plot. Professor Zhongliang Huang, Senior Engineer Binhui Chen, Associate Professor Juyu Lian, Yujia Huang, Dingshen Mo, Zhe Meng, Youchang Zhang, Linfang Wu, Yushi Ye ,Yinjie Chen, Wenbo Cai, Xiaoming Fang, Chuanyin Xiang and postgraduates Zhigao Wang, Junhui Shi, Chi Zhang, Lin Li, Shiguan Wei, Hao Shen, Jing Li, Hongpin Mu, Yujie Han, Guojun Lin, Wenping Liu, Linjuan Liao, Xiaoyi Li, Lei Ma, Young Shen, Shujun Xu, Bowen Li, Lanying Wang, Hongyu Niu, Ran Wang and Lei Dong in South China Botanical Garden of CAS, Dr. Zhengmei Lin in South China Teachers Training University, Fang Ten in Gutianshan Nature Reserve of Zhejiang province, undergraduate students Yuemou Shi, Dafeng Chen, Tenfang Xie, Ting Huang, Yuxi Chen, Shengqiang Li and Shuilian Chen in South China Teachers Training University and South China Agricultural University, and graduate students Shulan He, Xinda Li in Guangzhou University of Chinese Medicine Guohui Liang, Zhaoji Liu, Jingfeng Gu in South China Agricultural University , Yuanxiang Wen, Aqin Guo, Fenglin Pang, Guoqiang Yang in South China Teachers Training University worked hard during the establishment of the plot, the first and the second surveys of plants and the data collection. Linfang Wu, Yushi Ye, Zhengm Lin & Anqiang Dong provided us with the beautiful photos in this book. Professor Huagu Ye in South China Botanical Garden gave us valuable advice on the writing of this book.

In this book, we describes a total of 195 woody species including some varieties, belonging to 55 families and 111 genera. We identified and arranged gymnosperms and angiosperms according to the systems of Wanjun Zheng (1975) and Hutchinson, respectively. We described the biological characteristics of plants on the basis of *Flora of China* (Chinese and English version).

Due to the limitation of our time and knowledge, there might be unavoidable mistakes and oversights in this book, and we welcome comments and suggestions from our readers.

Ye Wanhui & Cao Honglin

2011.12

目 录

Contents

鼎湖山国家级自然保护区
Introduction to Dinghushan National Nature Reserve

1.1 地理位置和自然环境

中国科学院鼎湖山国家级自然保护区位于广东省肇庆市鼎湖区，地理坐标为北纬23° 09′ 21″ ～23° 11′ 30″，东经112° 30′ 39″ ～112° 33′ 41″ 。面积1155hm²，属低山丘陵地貌。最高峰鸡笼山海拔1000.3m，山体陡峭，坡度多在35° ～45° 之间。

本区属南亚热带湿润季风气候，冬夏气候交替明显。年平均温度20.9℃，最热月（7月）平均温度28.0℃，极端最高温度为 38.0℃，最冷月（1月）平均温度12.6℃。霜冻平均每年4次，每次持续 1～2天，极端最低温度 -0.2℃。1975～1995年年均降雨1985mm，4～9 月为主要降雨季节，月降雨量均超过200mm，11月至翌年1月为旱季，月降雨量不足100mm。年平均蒸发量 1115mm，年平均相对湿度80.3%。灾害性天气是寒潮和台风，寒潮常出现在 11月至翌年3月，年平均约3次，寒潮时气温迅速下降，并偶有霜冻。7～9 月为台风季节，平均每年有 4 次到达本地区。

区内地带性土壤为发育于砂岩和砂页岩的赤红壤，山地垂直分布有黄壤和山地灌丛草甸土。赤红壤分布于海拔 300m以下的丘陵低山，黄壤分布于海拔300～900m，900m以上为山地灌丛草甸土。此外，尚有局部分布的耕型赤红壤。

1.1 Location and description of Dinghushan National Nature Reserve

Dinghushan Nature Reserve, Chinese Academy of Sciences is located in the mid-part of Guangdong Province in South China, northeastern suburb of Zhaoqing city, about 84 km away from Guangzhou, with the geolocation of 112° 30′ 39″ -112° 33′ 41″ E and 23° 09′ 21″ -23° 11′ 30″ N. It occupies 1155 hm², covered mostly by hills and valleys. The altitude of the station ranges nearly from 100 to 700 m above sea level, with the highest of 1000.3 m at Jilongshan.

This area shares the typical monsoon climate, with the temperature alternate markedly between summer and winter. The annual mean temperature is 20.9℃, with the lowest of 12.6℃ in January and the highest of 28.0℃ in July. The frosts average 4 times a year which lasting out 1-2 days at a time. From 1975 to 1995 in the region the annual average precipitation was 1985 mm, the mean relative humidity was 80.3% and the mean amount of evaporation was 1115 mm. In summer, monthly rainfall is over 200 mm from April to September, whereas in winter, from November to January, monthly rainfall is less than 100 mm. Cold wave and typhoon are severe weather, with 3 cold waves from November to March and 4 typhoons from July to September.

The soil in Dinghushan is composed mainly of lateritic red earth and mountain yellow-brown earth in vertical distribution. The lateritic red earth occurs in hilly land below an altitude of 300 m, and also distributes in hills and low mountains at an altitude of 300 to 900 m above sea level, the mountain yellow-brown earth occurs partially on the tops of hills.

1.2 地带性植被类型

鼎湖山国家级自然保护区的地带性植被类型为南亚热带常绿阔叶林，主要分布于以庆云寺为中心的周围海拔75～500m的山坡，组成种类复杂多样且热带性植物较多。海拔500～800m的地段，分布着山地常绿阔叶林或灌丛林类型植被。海拔800m以上的山脊、山顶部分，只分布着灌木草丛类型植被。在海拔50～300m的沟谷中还分布着小面积的沟谷常绿阔叶林。在南亚热带常绿阔叶林的外围丘陵山地上，分布着较大面积的处于演替各阶段的针阔叶混交林。主要类型包括：

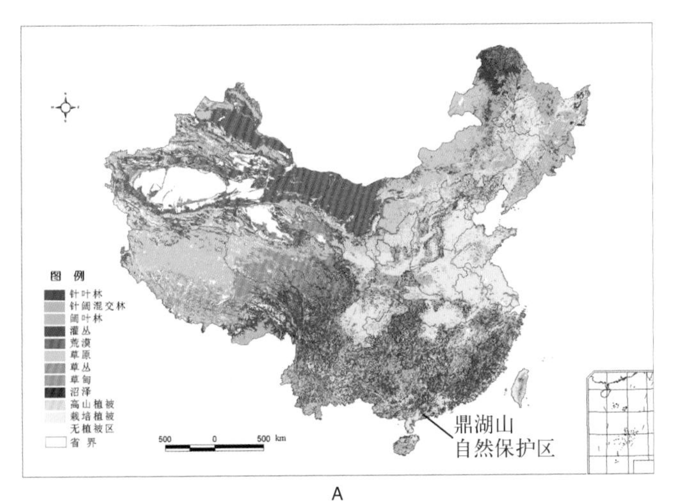

A

B

鼎湖山自然保护区的位置及20 hm²样地在鼎湖山保护区中的位置

Map of China showing the location of Dinghushan Nature Reserve and the 20 hm² Dinghushan permanent plot, southern China

A. 鼎湖山自然保护区的位置 The location of Dinghushan Nature Reserve in China

B. 20 hm²样地在鼎湖山保护区中的位置 The location of 20 hm² forest plot in Dinghushan Nature Reserve

1.2.1 常绿阔叶林

I. 南亚热带常绿阔叶林

1. 锥—黄果厚壳桂—云南银柴群落
2. 格木—黄果厚壳桂—鼎湖钓樟群落
3. 锥—黄果厚壳桂—黑桫椤群落
4. 木荷—锥—广东假木荷群落

II. 常绿阔叶林

5. 少叶黄杞—密花树—短序润楠群落

III. 沟谷常绿阔叶林

6. 九丁榕—青果榕—鱼尾葵群落
7. 橄榄—肥荚红豆—鱼尾葵群落

IV. 溪边林

8. 水翁蒲桃—蒲桃群落

1.2.2 针叶、阔叶混交林

V. 马尾松—阔叶混交林

9. 木荷—锥—马尾松群落
10. 广东润楠—鼠刺—马尾松群落
11. 木荷—岭南山竹子—马尾松群落
12. 木荷—马尾松—杜鹃花群落
13. 马尾松—木荷群落
14. 马尾松—木荷—桉树群落

1.2.3 热性针叶林

VI. 热性常绿针叶林

15. 马尾松—桃金娘—芒萁群落

1.2 Zonation of vegetation

Owing to its particular geographic location and ideal weather condition, Dinghushan Nature Reserve has rich vegetation types and is abundant in biological diversity. Vegetations can be divided into pine forest, mixed pine and broadleaved forest and south subtropical broadleaved evergreen forest along the horizontal level, and river-bank forest, ravine broadleaved evergreen forest, montane evergreen broadleaved forest, shrub-grasslands along the increasing vertical gradient. The seminatural forests include evergreen broadleaved forest, coniferous and broadleaved forest. The following are main vegetation types:

1.2.1 Broadleaved evergreen forest

I. South subtropical broadleaved evergreen forest

1. *Castanopsis chinensis* - *Cryptocarya concinna*- *Aporosa yunnanensis* Association
2. *Erythrophleum fordii* - *Cryptocarya concinna* - *Lindera chunii* Association
3. *Castanopsis chinensis* - *Cryptocarya concinna*- *Alsophila podophylla* Association
4. *Schima superba* - *Castanopsis chinensis*- *Craibiodendron scleranthum* var. *kwangtungense* Association

II. Broadleaved evergreen forest

5. *Engelhardtia fengelii* - *Myrsine seguinii* - *Machilus breviflora* Association

III. Ravine broadleaved evergreen forest

6. *Ficus nervosa* - *Ficus chlorocarpa* - *Caryota maxima* Association

7. *Canarium album* - *Ormosia fordiana* - *Caryota maxima* Association

IV. River-banks forest

8. *Syzygium nervosum* - *Syzygium jambos* Association

1.2.2 Coniferous and broadleaved mixed forest

V. Mix pine and broadleaved mixed forest

9. *Schima superba* - *Castanopsis chinensis* - *Pinus massoniana* Association

10. *Machilus kwangtungensis* - *Itea chinensis* - *Pinus massoniana* Association

11. *Schima superba* - *Garcinia oblongifolia* - *Pinus massoniana* Association

12. *Schima superba* - *Pinus massoniana* - *Rhododendron* spp. Association

13. *Pinus massoniana* - *Schima superba* Association

14. *Pinus massoniana* - *Schima superba* - *Eucalyptus* spp. Association

1.2.3 Tropical coniferous forest

VI. Tropical evergreen coniferous forest

15. *Pinus massoniana* - *Rhodomyrtus tomentosa-Dicranopteris pedata* Association

鼎湖山自然保护区不同的森林类型 Different types of forests in Dinghushan Nature Reserve

A. 南亚热带常绿阔叶林 South subtropical broadleaved evergreen forest

B. 沟谷常绿阔叶林 Ravine broadleaved evergreen forest

C. 针阔混交林 Pine and broadleaved trees mixed forest

D. 热带常绿针叶林 Tropical evergreen coniferous forest

2

鼎湖山南亚热带常绿阔叶林20hm²森林样地

The 20 hm² Lower Subtropical Broadleaved Evergreen Forest Plot in Dinghushan

2.1 样地建设与群落调查

样地植被为典型的南亚热带常绿阔叶林，位于鼎湖山国家级自然保护区内。样地面积为$20hm^2$，东西长400m，南北长500m，位于保护区中心地带。参照巴拿马巴洛科罗拉多岛（Barro Colorado Island, BCI）$50hm^2$热带雨林样地的技术规范，采用中国森林生物多样性监测网络的统一调查研究方法，于2005年在鼎湖山南亚热带常绿阔叶林建立了$20hm^2$ 固定监测样地，调查并鉴定了样方内胸径（diameter at breast height, DBH）大于1cm的木本植物，内容包括植物个体的物种名称、胸径、位置等，并挂牌标记。

2.1 Plot establishment and plant census

The forests in Dinghushan plot is representative of lower subtropical evergreen broadleaved forest in Dinghushan National Nature Reserve. A 20 hm^2 permanent plot of 400 m × 500 m was established in 2005 for long-term monitoring of biodiversity in the forest. The plot was established following the field protocol of the 50 hm^2 plot in Barro Colorado Island (BCI) in Panama. All free-standing trees with diameter at breast height (DBH) at least 1 cm were tagged, identified, measured, and georeferenced, corresponding data such as species, DBH, location data of each tagged tree was recorded.

2.2 地形和土壤

鼎湖山样地海拔230～470 m，坡度30° ～50°，地形起伏较大。样地土壤以发育于砂页岩母质的赤红壤为主，海拔300m以上的局部地段分布着山地黄壤。

2.2 Topography and soil

The DHS plot is characterized by rugged terrain, with altitude varying from 230 m to 470 m above sea level, and slope ranges from 30° to 50° . The soil in DHS plot is composed mainly of lateritic red earth below an altitude of 300 m, and mountain yellow-brown earth partially on the tops of hills.

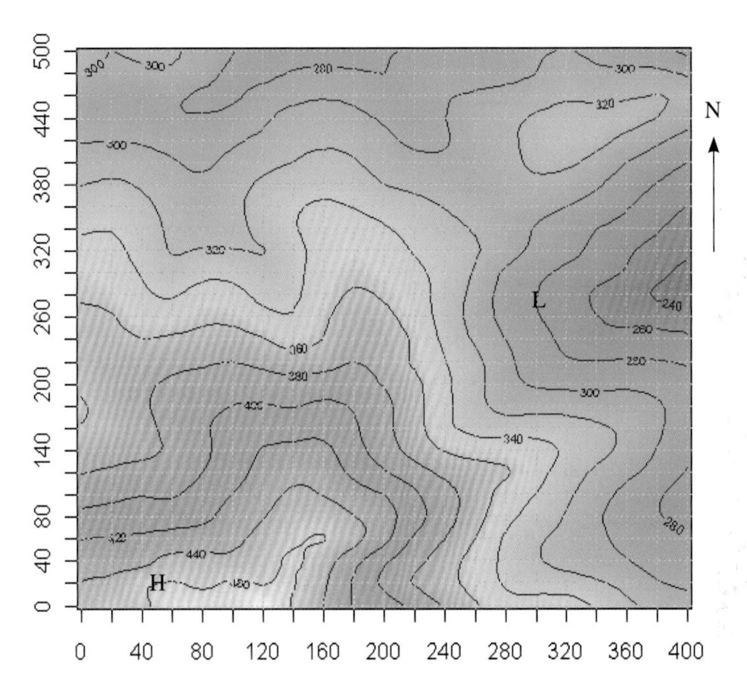

鼎湖山样地等高线图 The topography map of Dinghushan (DHS) plot

H: 最高点为470 m The highest = 470 m；L: 最低点为230 m The lowest = 230 m

2.3 物种组成和群落结构

样地内共有木本植物195种，71462个活的个体，分属于55科111属。从乔木区系的组成及其特点可以看出，其热带区系成分占绝对优势，并呈现出由亚热带向热带过渡的特色。

群落垂直结构复杂，地上成层现象较明显，乔木可分为3层，其中重要值最大的锥（*Castanopsis chinensis*）、木荷（*Schima superba*）和黄杞（*Engelhardtia roxburghiana*）均是乔木上层的优势种；中层是群落的主要层，由厚壳桂（*Cryptocarya chinensis*）、黄叶树（*Xanthophyllum hainanense*）和华润楠（*Machilus chinensis*）等中生和耐阴树种组成；下层成分较复杂，物种多样性高，不同地段的物种组成差异较大。

样地内物种十分丰富，种面积曲线拟合显示其物种数量接近于BCI。稀少种（指样地内个体数小于每公顷一株的种）比例极高，有98种，占总物种数的50.26%，其中有45%的稀少种源于物种本身的特性，有20%源于区系交汇，人为或自然干扰造成的稀少种占30%以上。

样地中所有个体的径级分布（以1 cm等级排列）明显呈倒“J”形，表示群落稳定并处于正常生长状态。根据对优势种径级结构的分布分析，将各树种的径级结构归纳为4种类型：（1）峰形（中径级个体储备型），此类物种为乔木上层优势种；（2）倒“J”形（正常型），此类物种为乔木中层的优势种；（3）类倒“J”形（偏正常型），此类物种占据乔木的中、下层；（4）“L”形（灌木型），此类物种分布于乔木下层和灌木层。

2.3 Species composition and community structure

There are 71462 individuals, belonging to 195 species, 111 genera and 55 families. Its floristic composition is transitional between the subtropical and tropical.

The vertical structure of the forest is clear. There are 5 layers from the top of the canopy to the ground floor, 3 tree layers (upper, middle and low), 1 shrub layer and 1 herb layer, respectively. Based on importance value, *Castanopsis chinensis*, *Schima superba* and *Engelhardtia roxburghiana* are the three most dominant species in the upper layer. There are many shade-tolerant and intermediate light-demanding species, such as *Cryptocarya chinensis*, *Xanthophyllum hainanense*, *Machilus chinensis* in mid-layer. Species in low layer are rich and complex, which

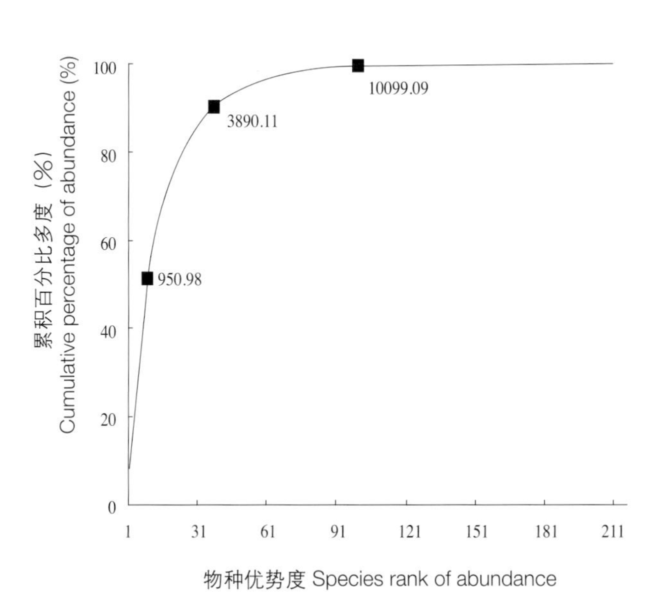

鼎湖山大样地物种多度的累计分布图
Cumulative distribution map of species abundance in Dinghushan (DHS) plot

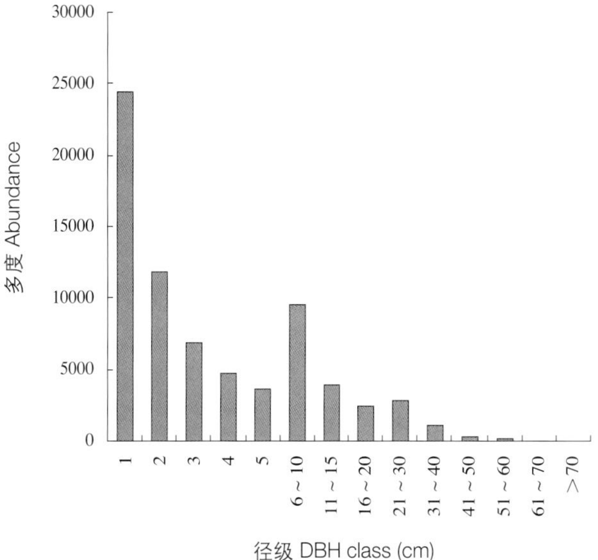

鼎湖山大样地木本植物径级分布图
Distribution patterns of size class of diameter at breast height (DBH) of tree species in Dinghushan（DHS）plot

composition varies a lot.

The species-area curve indicates that there is high diversity in the forest and the number of species is close to BCI. There is high proportion of rare species represented by 20% individuals which account for 50.26% of the total number of species. Among these rare species 45% of them lead to be rare by species characteristics, 20% by the floristic transitional nature of the plot, while the rest by disturbances.

Size distribution of all individuals shows an invert J-shape, which indicates that the community is in a stable and normal growth status. Size distributions of the dominant species are classified into four types based on their size-class frequencies, (1) unimodal in the top layer, (2) inverse J-shape in middle layer, (3) close to inverse J-shape in middle and low layer, (4) L-shape in low and shrub layer.

3

$20hm^2$ 南亚热带常绿阔叶林样地物种：木本植物及其分布格局

The 20 hm^2 Lower Subtropical Broadleaved Evergreen Forest Plot: Woody Plants and Their Distributions

1 长叶竹柏

chángyèzhúbǎi | Kim Giao

Nageia fleuryi (Hickel) de Laub.
罗汉松科 | Podocarpaceae

代码（SpCode）= NAGFLE
个体数（Individual number/20 hm^2）= 18
最大胸径（Max DBH）= 2.1 cm
重要值排序（Importance value rank）= 151

常绿乔木，高达30m，胸径可达70cm。树皮褐色、光滑，薄片状脱落。叶交互对生，厚革质，无中脉而为多数并列细脉，宽2.2～8cm，叶面深绿色，有光泽，叶背有多条气孔线。果球形，径1.5～1.8cm，熟时蓝紫色。花期3～4月，果10～11月成熟。

Evergreen trees to 30 m tall, trunk to 70 cm d.b.h.. Bark brownish purple, smooth, peeling in thin flakes. Leaves opposite, decussate, 2.2-8 cm, thickly leathery, stomatal lines present on abaxial surface only. Seed globose, 1.5-1.8 cm in diam.. bule-black when mature. Fl. Mar.-Apr., Fr. Oct.-Nov..

果枝 Fruiting branch
摄影：吴林芳 Photo by: Wu Linfang

幼苗 Seedling
摄影：吴林芳 Photo by: Wu Linfang

花序 Inflorescence
摄影：吴林芳 Photo by: Wu Linfang

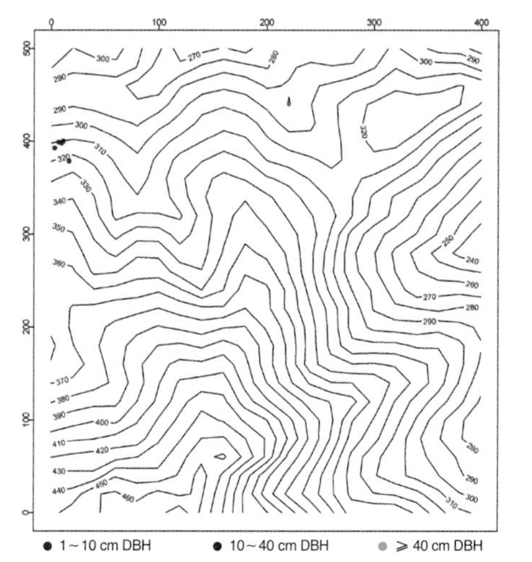

个体分布图 Distribution of individuals

径级分布表 DBH class

胸径等级 (Diameter class) (cm)	个体数 (No. of individuals in the plot)	比例 (Proportion) (%)
1～2	7	38.89
2～5	11	61.11
5～10	0	0.00
10～20	0	0.00
20～30	0	0.00
30～60	0	0.00
≥60	0	0.00

2 马尾松

mǎwěisōng | Masson Pine

Pinus massoniana Lambert
松科 I Pinaceae

代码（SpCode）= PINMAS
个体数（Individual number/20 hm^2）= 168
最大胸径（Max DBH）= 60.0 cm
重要值排序（Importance value rank）= 27

常绿乔木，高达45m。喜光，是干扰后裸地植被演替的先锋树种。树干皮红褐色至灰褐色，呈不规则开裂并片状剥落。枝条一般年生长两轮。针叶2针一束，罕3针1束。球果幼时绿色，成熟时栗褐色，卵形。花期4～5月，果翌年10～12月成熟。

Evergreen conifer, up to 45 m tall. Heliophilous, a pioneer species of the succession from denuded land after disturbance. Bark gray or red-brown, irregularly scaly and flaking. Branchlets usually growing twice per year. Needles 2 (or seldomly 3) per bundle. Seed cones green when young, turning chestnut brown at maturity, ovoid in shape. Fl. Apr.-May, fr. Oct. -Dec. of 2nd year.

树干 Trunk
摄影：吴林芳 Photo by: Wu Linfang

雄球花 Male cone
摄影：吴林芳 Photo by: Wu Linfang

球果 Cone
摄影：吴林芳 Photo by: Wu Linfang

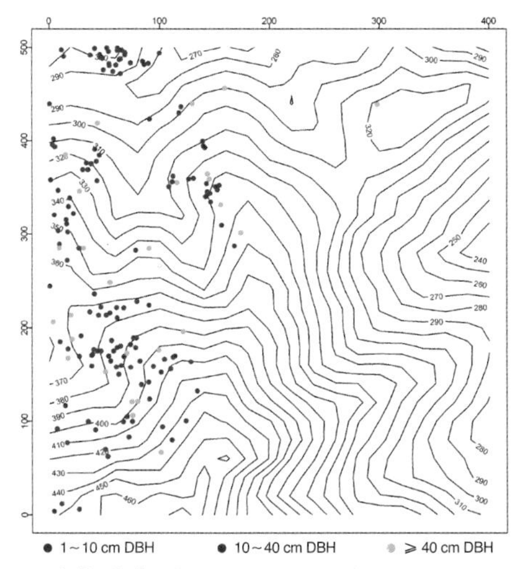

个体分布图 Distribution of individuals

径级分布表 DBH class

胸径等级 (Diameter class) (cm)	个体数 (No. of individuals in the plot)	比例 (Proportion) (%)
1～2	0	0.00
2～5	0	0.00
5～10	0	0.00
10～20	11	6.55
20～30	69	41.07
30～60	87	51.79
≥60	1	0.60

3 香港木兰（长叶玉兰） xiānggǎngmùlán | Hongkong Magnolia

Lirianthe championii (Benth.)N.H.Xia&C.Y.Wu
木兰科 | Magnoliaceae

代码（SpCode）= LIRCHA
个体数（Individual number/20 hm^2）= 11
最大胸径（Max DBH）= 4.9 cm
重要值排序（Importance value rank）= 119

常绿小乔木，高达11m。嫩枝绿色被毛。叶薄革质，椭圆形、狭长圆状椭圆形或狭倒卵状椭圆形，托叶痕几达叶柄顶端，边缘显著隆起。花球状，芳香，花被片9枚，外轮3枚，灰绿色，内轮6枚白色。聚合果褐色，椭圆形，长约3～4cm。花期4～6月，果期9～10月。

Evergreen small trees, up to 11 m tall. Twigs greenand hairy when young. Stipular scars nearly reached apex of petioles. Leaves elliptic, narrowly oblong-elliptic or narrowly obovate-elliptic, thinly leathery. Flowers globose, very fragrant, tepals 9, outer 3 pale green, inner 6 white. Aggregate fruit 3-4 cm, brown, ellipsoid. Fl. Apr.-Jun., fr. Sep.-Oct..

野外识别特征：

1. 叶面略皱如夜香木兰，但稍窄；
2. 芽、嫩枝、叶柄、叶中脉等被毛，而夜香木兰全株无毛；
3. 托叶痕几达叶柄顶端。

Key notes for identification:

1. Leaves ruqate as *coco*, but usually narrowly;
2. Buds, twigs, petioles and mid-vein hairy, but *coco* glabrous whole;
3. Stipular scars nearly reached apex of petioles.

花枝及果 Flowering branch and fruit
摄影：吴林芳 Photo by: Wu Linfang

叶 Leaf
摄影：吴林芳 Photo by: Wu Linfang

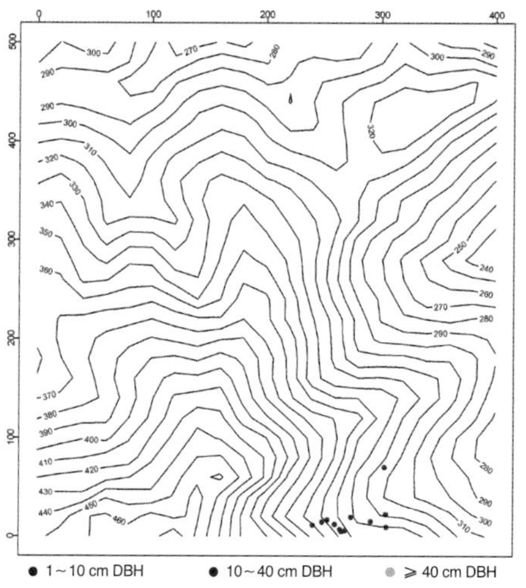

个体分布图 Distribution of individuals

径级分布表 DBH class

胸径等级 (Diameter class) (cm)	个体数 (No. of individuals in the plot)	比例 (Proportion) (%)
1～2	3	27.27
2～5	8	72.72
5～10	0	0.00
10～20	0	0.00
20～30	0	0.00
30～60	0	0.00
≥60	0	0.00

4 金叶含笑

jīnyèhánxiào | Goldenleaf Michelia

Michelia foveolata Merr.ex Dandy
木兰科 | Magnoliaceae

代码（SpCode）= MICFOV
个体数（Individual number/20 hm^2）= 1
最大胸径（Max DBH）= 60.2 cm
重要值排序（Importance value rank）= 120

常绿乔木，高30m。树皮淡灰或深灰色。芽、幼枝、叶柄、叶背、花梗密被红褐色、褐色或白色短茸毛。叶厚革质，叶基通常两侧不对称，叶面深绿色有光泽，叶背密被红铜色短茸毛，叶柄长1.5～3cm，无托叶痕。花被片9～12枚，淡黄绿色，基部带紫。聚合果长7～20cm，蓇葖果长圆状椭圆形。花期3～5月，果期9～10月。

Evergreen trees, to 30 m tall. Bark pale gray to dark gray. Young twigs, buds, petioles, leaf blade, abaxial surfaces, thickly leathery and brachyblasts densely reddish brown, brown, or white tomentulose. Petiole 1.5-3 cm, without a stipular scar. Leaf base usually asymmetrical. Tepals 9-12, pale yellowish green, base purplish. Aggregate fruit 7-20 cm, oblong-elliptic. Fl. Mar.-May, fr. Sep.-Oct..

野外识别特征：

1. 叶背密被红铜色短茸毛；
2. 花腋生；
3. 叶柄无托叶痕。

Key notes for identification:

1. Leaf abaxial surface densely reddish brown pubescence;
2. Flowers axillary;
3. Petiole without stipular scar.

花枝 Flowering branch
摄影：吴林芳 Photo by: Wu Linfang

果及叶 Fruit and leaf
摄影：吴林芳 Photo by: Wu Linfang

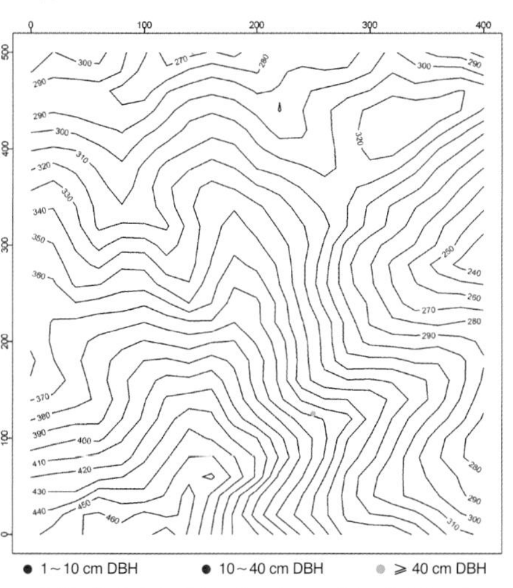

个体分布图 Distribution of individuals

径级分布表 DBH class

胸径等级 (Diameter class) (cm)	个体数 (No. of individuals in the plot)	比例 (Proportion) (%)
1～2	0	0.00
2～5	0	0.00
5～10	0	0.00
10～20	0	0.00
20～30	0	0.00
30～60	0	0.00
≥60	1	100.00

5 深山含笑

shēnshānhánxiào | Maud's Michelia

Michelia maudiae Dunn

木兰科 | Magnoliaceae

代码（SpCode）= MICMAU

个体数（Individual number/20 hm²）= 1

最大胸径（Max DBH）= 2.1 cm

重要值排序（Importance value rank）= 181

常绿乔木，高达20m。各部均无毛。树皮薄，浅灰色或灰褐色。芽、嫩枝、叶背和苞片均被白粉。叶革质，长圆状椭圆形，叶柄长1～3cm，无托叶痕。花芳香，花被片9枚，纯白色，基部稍呈淡红，聚合果长7～15cm。花期2～3月，果期9～10月成熟。

Evergreen trees, to 20 m tall. Glabrous. Bark pale gray or grayish brown, thin. Young twigs, buds, leaf abaxial surfaces, and bracts whitepowdery. Petiole 1-3 cm, without a stipular scar, leaf blade oblong-elliptic to rarely ovate-elliptic, leathery, Flowers fragrant, petals 9, white but base slightly pale red. Aggregate fruit 7-15 cm. Fl. Feb.-Mar., fr. Sep.-Oct..

野外识别特征：

1. 叶背灰绿色，被白粉；
2. 花腋生；
3. 叶柄无托叶痕。

Key notes for identification:

1. Leaf abaxial surface gray green, white powdery;
2. Flowers axillary;
3. Petiole without stipular scar.

花 Flower

摄影：吴林芳 Photo by: Wu Linfang

叶背及果 Leaf abaxial surface and fruit

摄影：吴林芳 Photo by: Wu Linfang

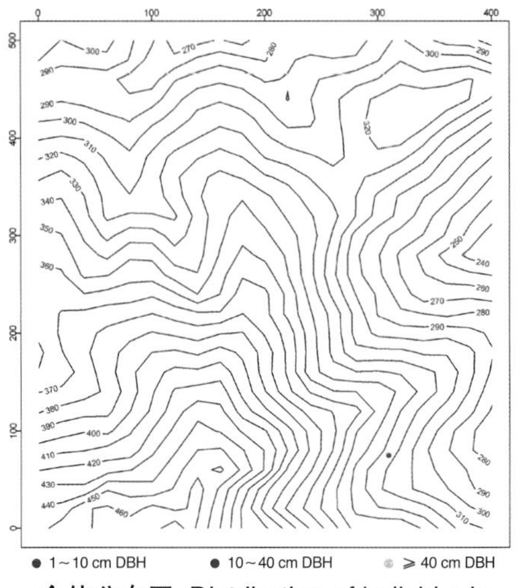

个体分布图 Distribution of individuals

径级分布表 DBH class

胸径等级 (Diameter class) (cm)	个体数 (No. of individuals in the plot)	比例 (Proportion) (%)
1～2	0	0.00
2～5	1	100.00
5～10	0	0.00
10～20	0	0.00
20～30	0	0.00
30～60	0	0.00
≥60	0	0.00

6 厚壳桂 hòukéguì | Chinese Cryptocarya

Cryptocarya chinensis (Hance) Hemsl.
樟科 | Lauraceae

代码（SpCode）= CRYCHI
个体数（Individual number/20 hm^2）= 2546
最大胸径（Max DBH）= 57.0 cm
重要值排序（Importance value rank）= 7

常绿乔木，高达20m，胸径可达50cm。树皮暗灰色，粗糙。叶互生或对生，革质，长椭圆形，叶面光亮叶背苍白，具离基三出脉，中脉叶面凹叶背凸，上部侧脉2～3对互生。圆锥花序腋生及顶生，花淡黄色。果球形或扁球形，紫黑色，具纵棱12～15条。花期4～5月，果期8～12月。

Evergreen trees, up to 20 m tall, to 50 cm d.b.h.. Bark dark gray, scabrid. Leaves alternate or opposite, leaf blade leathery, narrowly elliptic, triplinerved, midrib elevated abaxially, impressed adaxially, upper lateral veins 2 or 3 pairs and alternate. Panicles axillary and terminal, flowers yellowish. Fruit globose or oblate, purple-black, 12-15-angulate. Fl. Apr.-May, fr. Aug.-Dec..

野外识别特征：

1. 树皮不具深纵裂；
2. 叶具离基三出脉；
3. 叶背苍白色。

Key notes for identification:

1. Bark not longitudinally deeply fissured;
2. Leaf blade triplinerved;
3. Leaf blade abaxially pale.

果枝 Fruiting branch
摄影：吴林芳 Photo by: Wu Linfang

花枝 Flowering branch
摄影：吴林芳 Photo by: Wu Linfang

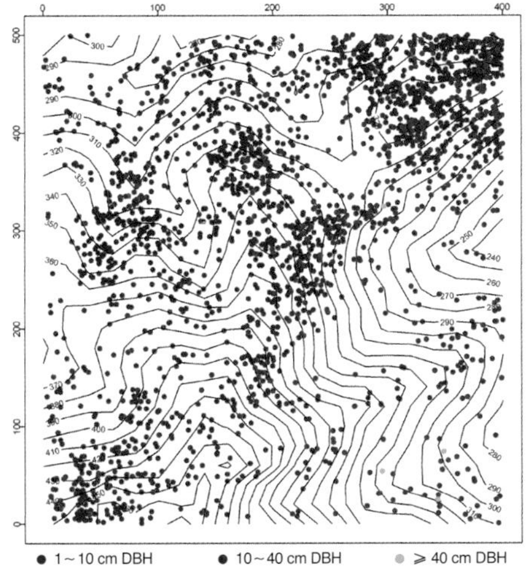

个体分布图 Distribution of individuals

径级分布表 DBH class

胸径等级 (Diameter class) (cm)	个体数 (No. of individuals in the plot)	比例 (Proportion) (%)
1～2	360	14.14
2～5	781	30.68
5～10	772	30.32
10～20	438	17.20
20～30	145	5.70
30～60	50	1.96
≥60	0	0.00

7 黄果厚壳桂（生虫树） huángguǒhòukéguì | Concinna Cryptocarya

Cryptocarya concinna Hance
樟科 | Lauraceae

代码（SpCode）= CRYCON
个体数（Individual number/20 hm²）= 4449
最大胸径（Max DBH）= 47.1 cm
重要值排序（Importance value rank）= 8

常绿乔木，高达18m，胸径达35cm。树皮淡褐色，常光滑。叶互生，椭圆状长圆形或长圆形，坚纸质，羽状脉，叶面不明显，叶背明显，中脉叶面凹叶背凸。圆锥花序腋生及顶生。果长椭圆形，幼时深绿，熟时黑或蓝黑。花期3～5月，果期6～12月。

Evergreen trees, up to 18 m tall, to 35 cm d.b.h.. Bark brownish, usually glabrous. Leaves alternate, leaf blade elliptic-oblong or oblong, stiffly papery, pinnate vein, slightly conspicuous abaxially, midrib elevated abaxially, impressed adaxially. Panicles axillary and terminal. Fruit dark green when young, black or blue-black when mature, narrowly ellipsoid, Fl. Mar.-May, fr. Jun.-Dec.

幼苗 Seedling
摄影：吴林芳 Photo by: Wu Linfang

果枝 Fruiting branch
摄影：吴林芳 Photo by: Wu Linfang

花枝 Flowering branch
摄影：吴林芳 Photo by: Wu Linfang

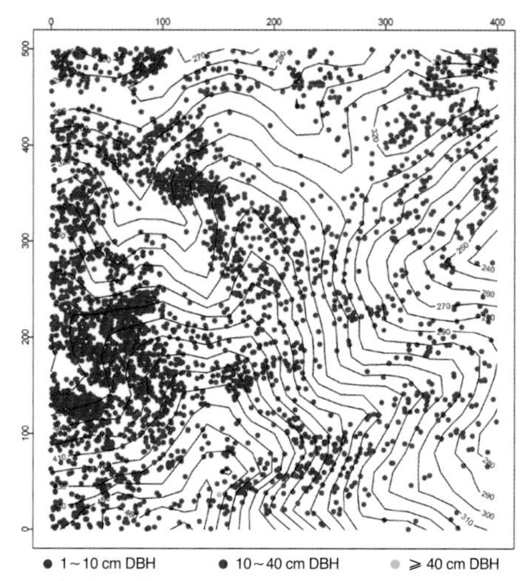

个体分布图 Distribution of individuals

径级分布表 DBH class

胸径等级 (Diameter class) (cm)	个体数 (No. of individuals in the plot)	比例 (Proportion) (%)
1～2	3845	86.42
2～5	466	10.47
5～10	41	0.92
10～20	81	1.82
20～30	15	0.34
30～60	1	0.02
≥60	0	0.00

8 乌药

wūyào | Combined Spicebush

Lindera aggregata (Sims) Kosterm.
樟科 | Lauraceae

代码（SpCode）= LINAGG
个体数（Individual number/20 hm^2）= 4
最大胸径（Max DBH）= 6 cm
重要值排序（Importance value rank）= 152

常绿灌木或小乔木。树皮灰褐色，幼枝青绿色，具纵向细纹，密被金黄色绢毛，后脱落。叶互生，卵形至近圆形，革质或近革质，三出脉，叶面绿色有光泽，叶背苍白，幼时被毛。伞形花序腋生，无总梗，常6～8朵聚生极短枝上。果卵形或近球形，长6～10mm，径4～7mm。花期3～4月，果期5～9月。

Evergreen shrubs or small trees. Bark grayish brown, young branchlets blue-green, longitudinally striate, densely golden sericeous later gradually deciduous and glabrous. Leaves alternate, ovate to subrounded, leathery or subleathery, trinerved, blade pale abaxially, green and shiny and pubescent when young adaxially. Umbels 6-8, inserted in short branchlet, axillary, not pedunculate. Fruits ovate or sometimes subrounded, 6-10 mm × 4-7 mm. Fl. Mar.-Apr., fr. May-Nov..

野外识别特征：

1. 叶三出脉，革质或近革质；
2. 叶尾尖；
3. 叶面绿色有光泽，叶背苍白色。

Key notes for identification:

1. Leaf blade trinerved, leathery or subleathery;
2. Leaf blade apex acuminate or caudate-acuminate;
3. Leaf blade pale abaxially, green and shiny and pubescent when young adaxially.

果枝 Fruiting branch
摄影：吴林芳 Photo by: Wu Linfang

花 Flower
摄影：吴林芳 Photo by: Wu Linfang

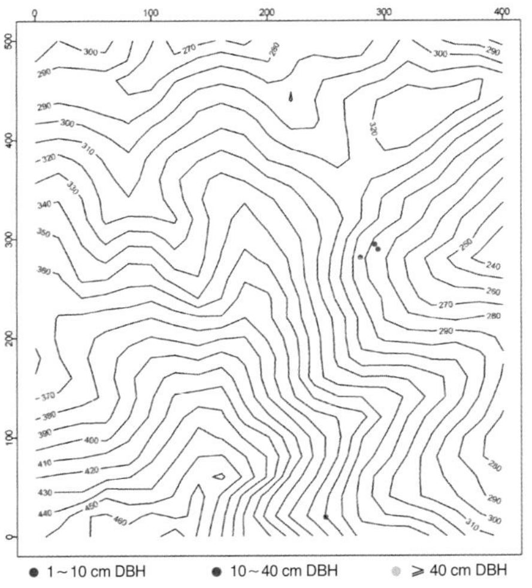

个体分布图 Distribution of individuals

径级分布表 DBH class

胸径等级 (Diameter class) (cm)	个体数 (No. of individuals in the plot)	比例 (Proportion) (%)
1～2	0	0.00
2～5	2	50.00
5～10	2	50.00
10～20	0	0.00
20～30	0	0.00
30～60	0	0.00
≥60	0	0.00

9 鼎湖钓樟

dǐnghúdiàozhāng | Chun's Spicebush

Lindera chunii Merr.

樟科 | Lauraceae

代码（SpCode）= LINCHU

个体数（Individual number/20 hm²）= 1293

最大胸径（Max DBH）= 16.0 cm

重要值排序（Importance value rank）= 19

常绿灌木或小乔木，高6m。叶互生，椭圆形至长椭圆形，纸质，幼时两面被白毛或金黄色贴伏绢毛，三出脉，侧脉至先端。伞形花序数个生于叶腋短枝上，每伞形花序有花4～6朵。果椭圆形，长8～10mm，径6～7mm。花期2～3月，果期8～9月。

Evergreen shrubs or small trees, to 6 m tall. Leaves alternate, elliptic oblong, papery, white or golden and appressed sericeous on both surfaces when young, trinerved, lateral veins reaching apex. Umbels several inserted at short branch in leaf axil, 4-6-flowered. Fruits ellipsoid, 8-10 mm × 6-7 mm. Fl. Feb.-Mar., fr. Aug.-Sep..

野外识别特征：

1. 叶基三出脉，侧脉至先端，纸质；
2. 叶尾尖；
3. 叶面黄绿色，叶背被红褐色鳞毛。

Key notes for identification:

1. Leaf blade trinerved, lateral veins reaching apex, papery;
2. Leaf blade apex caudate-acuminate;
3. Leaf blade yellowgreen adaxially, red brown scalyhair abaxially.

叶 Leaf

摄影：吴林芳 Photo by: Wu Linfang

果枝及花 Fruiting branch & flower

摄影：吴林芳 Photo by: Wu Linfang

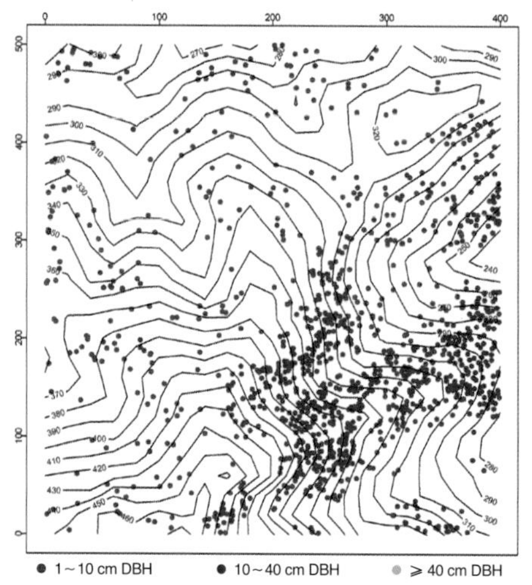

个体分布图 Distribution of individuals

径级分布表 DBH class

胸径等级 (Diameter class) (cm)	个体数 (No. of individuals in the plot)	比例 (Proportion) (%)
1～2	238	18.41
2～5	690	53.36
5～10	339	26.22
10～20	26	2.01
20～30	0	0.00
30～60	0	0.00
≥60	0	0.00

10 香叶树 xiāngyèshù | Chinese Spicebush

Lindera communis Hemsl.
樟科 | Lauraceae

代码（SpCode）= LINCOM
个体数（Individual number/20 hm^2）= 1
最大胸径（Max DBH）= 1.2 cm
重要值排序（Importance value rank）= 192

常绿灌木至小乔，高达15m，胸径达25cm。树皮淡褐色。叶互生，披针形、卵形或椭圆形，革质，叶面绿色，无毛，背面灰绿或浅黄色，被黄褐色柔毛后脱落，边缘内卷，羽状脉。伞形花序，单生或二个同生于叶腋，总梗极短。果卵形至球形，熟时红色。花期3～4月，果期9～10月。

Evergreen shrubs or trees, to 15 m tall, 25 cm d.b.h. Bark brownish. Leaves alternate, leaf blade gray-green or yellowish abaxially, green adaxially, lanceolate, ovate, or elliptic, leathery, yellow-brown pubescent, later laxly pubescent, margin involute, pinninerved. Umbels solitary or 2, inserted in leaf axil; peduncles very short. Fruit ovate, red at maturity. Fl. Mar.-Apr., fr. Sep.-Oct..

野外识别特征：
1. 树皮淡褐色，常光滑；
2. 叶羽状脉，革质；
3. 伞形花序几无梗或梗极短。

Key notes for identification:
1. Bark brownish, usually smooth;
2. Leaf blade pinninerved, leathery;
3. Umbels no peduncles or peduncles very short.

叶 Leaf
摄影：吴林芳 Photo by: Wu Linfang

果枝及花 Fruiting branch and flower
摄影：吴林芳 Photo by: Wu Linfang

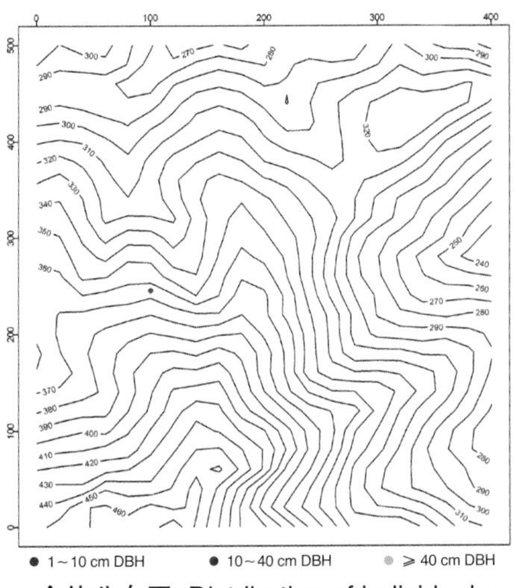

个体分布图 Distribution of individuals

径级分布表 DBH class

胸径等级 (Diameter class) (cm)	个体数 (No. of individuals in the plot)	比例 (Proportion) (%)
1～2	1	100.00
2～5	0	0.00
5～10	0	0.00
10～20	0	0.00
20～30	0	0.00
30～60	0	0.00
≥60	0	0.00

11 滇粤山胡椒（山钓樟） diānyuèshānhújiāo | Metcalfiana Spicebush

Lindera metcalfiana Allen
樟科 | Lauraceae

代码（SpCode）= LINMET
个体数（Individual number/20 hm^2）= 2113
最大胸径（Max DBH）= 30.0 cm
重要值排序（Importance value rank）= 16

落叶小乔木，高3～12m。树皮灰黑或淡褐色。枝纤细，圆柱形，幼时多少具棱，初时略被黄褐色绢质微柔毛，后渐变无毛。叶纸质，互生，椭圆形或长椭圆形，羽状脉。花黄色，分雌雄聚伞花序，1～2（3）个生于叶腋内短枝上。果熟时为紫黑色，球形，径6mm。花期3～5月，果期6～10月。

Deciduous small trees, 3-12 m tall. Bark gray-black or brownish. Branchlets slender, terete and ± angular when young, laxly yellow-brown sericeous-pubescent, later glabrate. Leaves alternate, leaf blade elliptic, narrowly elliptic, or lanceolate, papery, pinninerved. Flowers yellow, male ang female umbels 1 or 2 (or 3) , inserted at axillary short branch. Fruits globose, ca. 6 mm in diam., purple-black at maturity. Fl. Mar.-May, fr. Jun.-Oct..

野外识别特征：
1. 叶羽状脉；
2. 叶纸质；
3. 花序具明显梗；
4. 幼树皮灰褐色。

Key notes for identification:
1. Leaf blade pinninerved;
2. Leaf blade papery;
3. Umbels with evident peduncles;
4. Bark gray-brown when young.

果枝 Fruiting branch
摄影：吴林芳 Photo by: Wu Linfang

叶 Leaf
摄影：吴林芳 Photo by: Wu Linfang

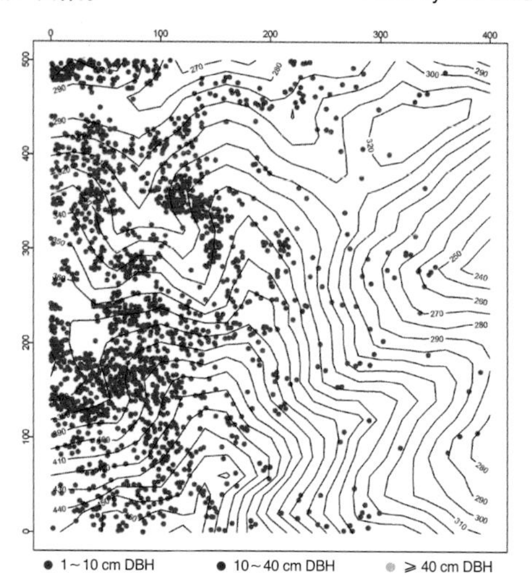

个体分布图 Distribution of individuals

径级分布表 DBH class

胸径等级 (Diameter class) (cm)	个体数 (No. of individuals in the plot)	比例 (Proportion) (%)
1～2	622	29.44
2～5	1005	47.56
5～10	418	19.78
10～20	67	3.17
20～30	1	0.05
30～60	0	0.00
≥60	0	0.00

12 山鸡椒（山苍子） shānjījiāo | Fragrant Litsea

Litsea cubeba (Lour.) Per.
樟科 | Lauraceae

代码（SpCode）= LITCUB
个体数（Individual number/20 hm²）= 51
最大胸径（Max DBH）= 10.7 cm
重要值排序（Importance value rank）= 87

落叶小乔木，高8～10m。幼树皮黄绿色，光滑，老树皮灰褐色。小枝无毛。叶互生，披针形或长圆形，纸质，羽状脉，两面无毛。伞房花序有花4～6朵，单生或簇生，先叶开放或与叶同时开放。果近球形，径5mm。花期2～4月，果期5～8月。

Deciduous small trees, 8-10 m tall. Bark yellow-green and smooth when young, gray-brown when old. Branchlets glabrous. Leaves alternate, glabrous, leaf blade lanceolate, oblong, or elliptic, papery, pinninerved, glabrous on both surfaces. Umbels solitary or clustered, 4-6-flowered, flowering before leaves or with leaves. Fruit subglobose, 5 mm in diam.. Fl. Feb.-Apr., fr. May-Aug..

野外识别特征：

1. 叶羽状脉，纸质；
2. 叶披针形或长圆形；
3. 幼树皮黄绿色，光滑。

Key notes for identification:

1. Leaf blade pinninerved, papery;
2. Leaf blade lanceolate, oblong, or elliptic;
3. Bark yellow-green and smooth when young.

果枝 Fruiting branch
摄影：吴林芳 Photo by: Wu Linfang

花序 Inflorescence
摄影：吴林芳 Photo by: Wu Linfang

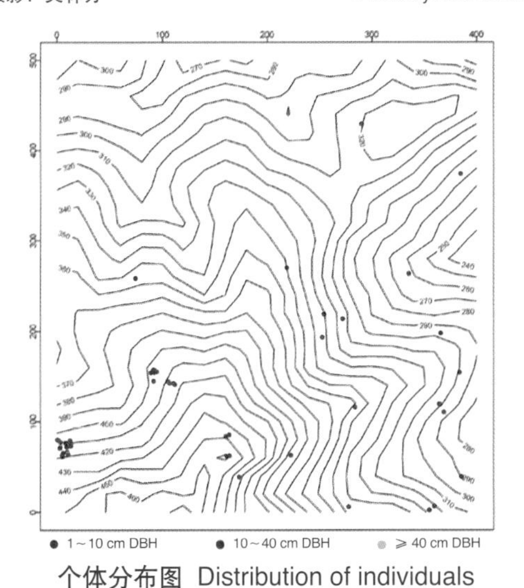

个体分布图 Distribution of individuals

径级分布表 DBH class

胸径等级 (Diameter class) (cm)	个体数 (No. of individuals in the plot)	比例 (Proportion) (%)
1～2	17	33.33
2～5	28	54.90
5～10	4	7.84
10～20	2	3.92
20～30	0	0.00
30～60	0	0.00
≥60	0	0.00

13 豺皮樟

cháipízhāng | Oblong-leaved Litsea

Litsea rotundifolia Hemsl. var. *oblongifolia* (Nees) Allen
樟科 | Lauraceae

代码（SpCode）= LITROT
个体数（Individual number/20 hm^2）= 8
最大胸径（Max DBH）= 6.6 cm
重要值排序（Importance value rank）= 137

常绿灌木或小乔木，高可达5m。树皮灰色或灰褐色。叶散生，革质，卵状长圆形，两面无毛，叶背粉绿色，羽状脉。伞房花序常3个簇生于叶腋，几无总梗，每花序有花3～4朵。果球形，径6mm，熟时灰蓝黑色。花期8～9月，果期9～11月。

Evergreen shrubs or small trees, up to 5 m tall. Bark gray or gray-brown. Leaves scattered, leaf blade ovate-oblong, glabrous on both surfaces and glaucous abaxially, pinninerved, leathery. Umbels often in cluster of 3, axillary, almost sessile, 3- or 4-flowered. Fruit globose, 6 mm in diam., gray-blue-black at maturity. Fl. Aug.-Sep., fr. Sep.-Nov..

野外识别特征：

1. 叶羽状脉，革质；
2. 叶卵状长圆形，叶背粉绿色；
3. 树皮常具褐色斑点。

Key notes for identification:

1. Leaf blade pinninerved, leathery;
2. Leaf blade ovate-oblong, glaucous abaxially;
3. Bark usually with brown patches.

花枝 Flowering branch
摄影：吴林芳 Photo by: Wu Linfang

叶及果枝 Leaf and fruiting branch
摄影：吴林芳 Photo by: Wu Linfang

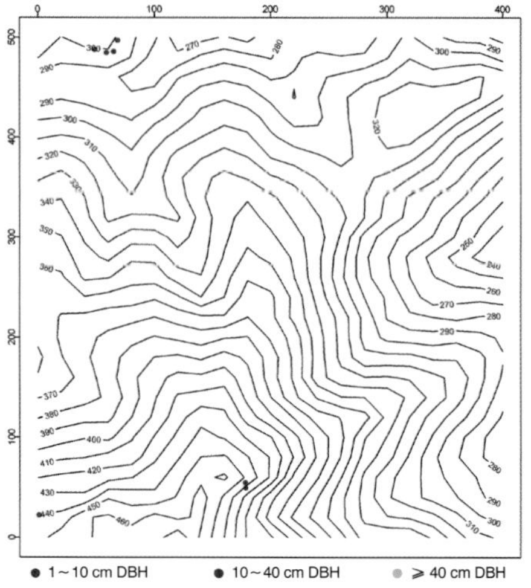

个体分布图 Distribution of individuals

径级分布表 DBH class

胸径等级 (Diameter class) (cm)	个体数 (No. of individuals in the plot)	比例 (Proportion) (%)
1～2	1	12.50
2～5	6	75.00
5～10	1	12.50
10～20	0	0.00
20～30	0	0.00
30～60	0	0.00
≥60	0	0.00

14 轮叶木姜子

lúnyèmùjiāngzǐ | Verticillate Litsea

Litsea verticillata Hance
樟科 | Lauraceae

代码（SpCode）= LITVER
个体数（Individual number/20 hm^2）= 196
最大胸径（Max DBH）= 11.6 cm
重要值排序（Importance value rank）= 57

常绿灌木或小乔木，高可达5m。小枝，密被黄色长硬毛，无毛。叶4～6片轮生，披针形或倒披针状长椭圆形，叶背被毛，羽状脉。伞房花序2～10个集生于小枝顶，每花序有花5～8朵，淡黄色，近无梗。果卵形或椭圆形。花期4～11月，果期11月至翌年1月。

Evergreen shrubs or small trees, to 5 m tall. Young branchlets densely yellow hirsute and becoming glabrous. Leaves 4-6-verticillate, densely yellow villous, leaf blade long lanceolate or long elliptic-oblanceolate, pubescent abaxially, pinninerved. Umbels in cluster of 2-10 at apex of branchlet, 5-8-flowered, pale yellow, subsessile. Fruit ovoid or ellipsoid. Fl. Apr.-Nov., fr. Nov.-Jan. of next year.

花 Flower
摄影：吴林芳 Photo by: Wu Linfang

叶 Leaf
摄影：吴林芳 Photo by: Wu Linfang

果 Fruit
摄影：吴林芳 Photo by: Wu Linfang

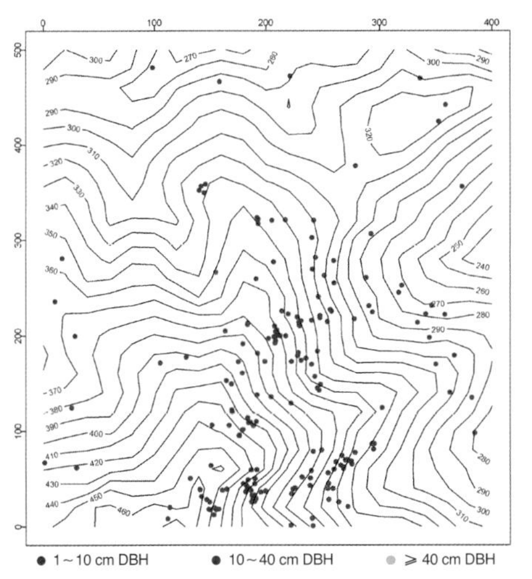

个体分布图 Distribution of individuals

径级分布表 DBH class

胸径等级 (Diameter class) (cm)	个体数 (No. of individuals in the plot)	比例 (Proportion) (%)
1～2	89	45.41
2～5	86	43.88
5～10	20	10.20
10～20	1	0.51
20～30	0	0.00
30～60	0	0.00
≥60	0	0.00

15 短序润楠 duǎnxùrùnnán | Short-flowered Machlius

Machilus breviflora (Benth.) Hemsl.
樟科 | Lauraceae

代码（SpCode）= MACBRE
个体数（Individual number/20 hm^2）= 844
最大胸径（Max DBH）= 41.5 cm
重要值排序（Importance value rank）= 21

常绿乔木，高约8m。树皮灰褐色。叶略生枝顶，倒卵形至倒卵状披针形，两面无毛，中脉叶面凹叶背凸。圆锥花序3～5个顶生，花绿白色。果球形，径8～10mm。花期7～8月，果期10～12月。

Evergreen trees, to 8 m tall. Bark gray-brown. Leaf blade obovate to obovate-lanceolate, glabrous on both surfaces, midrib raised abaxially, concave adaxially. Panicles terminal. Flowers green-white. Fruit globose, 8-10 mm in diam.. Fl. Jul.-Aug., fr. Oct.-Dec..

野外识别特征：

1. 叶常聚生枝顶；
2. 叶倒卵形至倒卵状披针形；
3. 叶一般较华润楠短，长约3～5cm；叶柄长3～5mm或更短；
4. 树皮灰褐色，具皮孔。

Key notes for identification:

1. Leaves usually consorte top of branchlets;
2. Leaf blade obovate to obovate-lanceolate;
3. Leaves usually shorter than *M. chinensis*, apex obtuse; length 3-5 cm, Petiole 3-5 mm or less;
4. Bark gray-brown, with lenticel.

果枝 Fruiting branch
摄影：吴林芳 Photo by: Wu Linfang

花枝 Flowering branch
摄影：吴林芳 Photo by: Wu Linfang

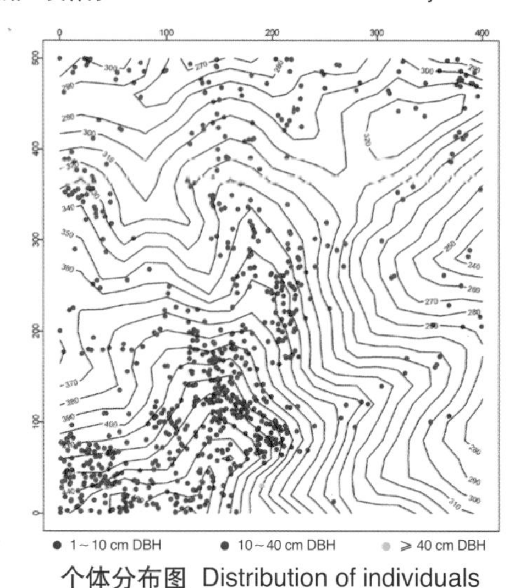

个体分布图 Distribution of individuals

径级分布表 DBH class

胸径等级 (Diameter class) (cm)	个体数 (No. of individuals in the plot)	比例 (Proportion) (%)
1～2	100	11.85
2～5	193	22.87
5～10	307	36.37
10～20	216	25.59
20～30	19	2.25
30～60	9	1.07
≥60	0	0.00

16 华润楠

huárùnnán | Chinese Machilus

Machilus chinensis (Benth.) Hemsl.
樟科 | Lauraceae

代码（SpCode）= MACCHI
个体数（Individual number/20 hm^2）= 507
最大胸径（Max DBH）= 63.0 cm
重要值排序（Importance value rank）= 17

常绿乔木，高可达20m。树皮灰色，全株无毛。叶倒卵状长椭圆形至长椭圆状倒披针形，革质，中脉叶面凹叶背凸，侧脉不明显。圆锥花序顶生，常较叶短，在上部分枝，每花序有花5～10朵，总梗占全长的3/4，花白色。果球形，径8～11mm。花期10～11月，果期12月至翌年2月。

Evergreen trees, to 20 m tall. Bark gray, complete stool glabrous. leaf blade obovate-oblong to oblong-oblanceolate, leathery, glabrous on both surfaces, midrib raised abaxially, concave adaxially. Panicles usually terminal, shorter than leaf blade, with 5-10 flowers, branched at upper part of peduncle, flowers white. Fruit globose, 8-11 mm in diam.. Fl. Oct.-Nov., fr. Dec. to Feb. of next year.

野外识别特征：

1. 叶倒卵状长椭圆形至长椭圆状披针形，先端钝或短渐尖；
2. 叶片长5～8（10）cm，叶柄长6～14mm；
3. 树皮灰色，具皮孔。

Key notes for identification:

1. Leaf blade obovate-oblong to oblong-oblanceolate, apex obtuse or shortly acuminate;
2. Leaf blade length 5-8 (10) cm, petiole 6-14 mm;
3. Bark gray, with lenticel.

果枝 Fruiting branch
摄影：吴林芳 Photo by: Wu Linfang

叶 Leaf
摄影：吴林芳 Photo by: Wu Linfang

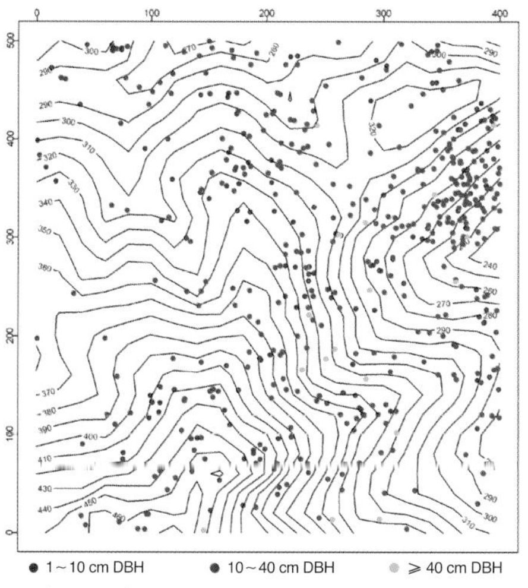

个体分布图 Distribution of individuals

径级分布表 DBH class

胸径等级 (Diameter class) (cm)	个体数 (No. of individuals in the plot)	比例 (Proportion) (%)
1～2	40	7.89
2～5	71	14.00
5～10	77	15.19
10～20	121	23.87
20～30	122	24.06
30～60	74	14.60
≥60	2	0.39

17 广东润楠

guǎngdōngrùnnán | Kwangtung Machilus

Machilus kwangtungensis Yang

樟科 | Lauraceae

代码（SpCode）= MACKWA

个体数（Individual number/20 hm²）= 12

最大胸径（Max DBH）= 38.0 cm

重要值排序（Importance value rank）= 104

常绿乔木，高达10m。当年生枝条密被锈色茸毛，1、2年生枝条无毛，有黄褐色纵裂唇形皮孔。叶长椭圆形或倒披针形，薄革质，叶面深绿无毛，叶背淡绿有贴伏小柔毛。圆锥花序生于新枝下端，有灰黄色小柔毛；花被裂片近等长，两面被小柔毛。果近球形，略扁，径8～9mm，熟时黑色。花期3～4月，果期5～7月。

Evergreen trees, to 10 m tall. Young shoots densely ferruginous tomentose, 1-year-old and 2-year-old branchlets glabrous, with yellowish brown lenticels. Leaf blade oblong or oblanceolate, thinly leathery, abaxially appressed pubescent, adaxially glabrate. Panicles arising from lower part of current year branchlet, with grayish yellow pubescent, perianth lobes subequal, grayish yellow pubescent on both surfaces. Fruit subglobose, slightly compressed, 8-9 mm in diam., blackish when ripe. Fl. Mar.-Apr., fr. May-Jul..

野外识别特征：

1. 当年生枝密被锈色茸毛；
2. 叶面深绿无毛，叶背淡绿有贴伏小柔毛；
3. 叶薄革质，先端短尖。

Key notes for identification:

1. Twigs densely ferruginous tomentose;
2. Leaf blade dark green glabrate adaxially, palegreen and appressed pubescent abaxially;
3. Leaf blade thinly leathery, apex acuminate.

果枝 Fruiting branch

摄影：吴林芳 Photo by: Wu Linfang

花序 Inflorescence

摄影：吴林芳 Photo by: Wu Linfang

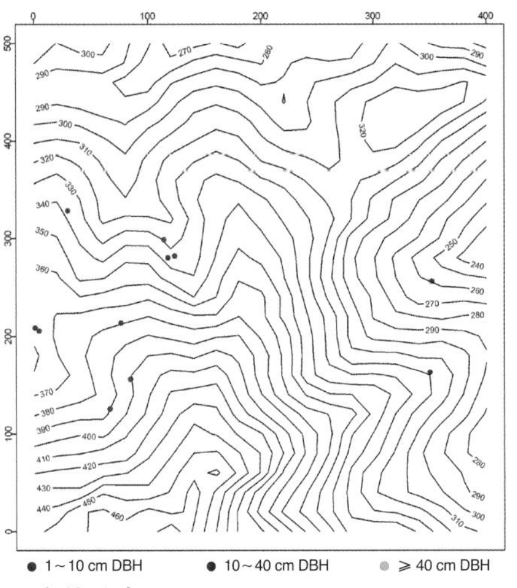

个体分布图 Distribution of individuals

径级分布表 DBH class

胸径等级 (Diameter class) (cm)	个体数 (No. of individuals in the plot)	比例 (Proportion) (%)
1～2	3	25.00
2～5	4	33.33
5～10	0	0.00
10～20	4	33.33
20～30	0	0.00
30～60	1	8.33
≥60	0	0.00

18 凤凰润楠 fènghuángrùnnán | Phoenix Machilus

Machilus phoenicis Dunn
樟科 | Lauraceae

代码（SpCode）= MACPHO
个体数（Individual number/20 hm^2）= 48
最大胸径（Max DBH）= 7.2 cm
重要值排序（Importance value rank）= 81

常绿中等乔木，高5m。树皮褐色，全株无毛。枝和小枝粗壮，紫褐色。叶椭圆形、长椭圆形至狭长椭圆形，厚革质，中脉下面粗壮，明显凸起。花序多数，生于枝端，长5～8cm，在上端分枝，略带红褐色。果球形，径约9mm，果梗增粗。花期3～5月，果期6～9月。

Evergreen medium-sized trees, to 5 m tall. Bark brown, glabrous throughout. Branchlets purple-brown, stout. Leaf blade elliptic or oblong to narrowly oblong, thickly leathery, midrib abaxially strongly conspicuously elevated, Panicles numerous, arising from apex of branchlet, 5-8 cm, branched at upper part of peduncle, peduncle reddish brown slightly. Fruit globose, ca. 9 mm in diam., fruiting pedicel enlarged. Fl. Mar.-May, fr. Jun.-Sep..

野外识别特征：

1. 树皮褐色，具纵深裂纹；
2. 枝和小枝粗壮，紫褐色；
3. 叶厚革质，叶缘略反卷；
4. 叶中脉叶背明显凸起，略带红褐色。

Key notes for identification:

1. Bark brown, longitudinally deeply fissured;
2. Branchlets sometimes purple-brown, stout;
3. Leaf blade thickly leathery, margin slightly revolute;
4. Midrib abaxially strongly conspicuously elevated, slightly reddish brown.

果枝 Fruiting branch
摄影：吴林芳 Photo by: Wu Linfang

叶背及树干 Leaf abaxial surface and trunk
摄影：吴林芳 Photo by: Wu Linfang

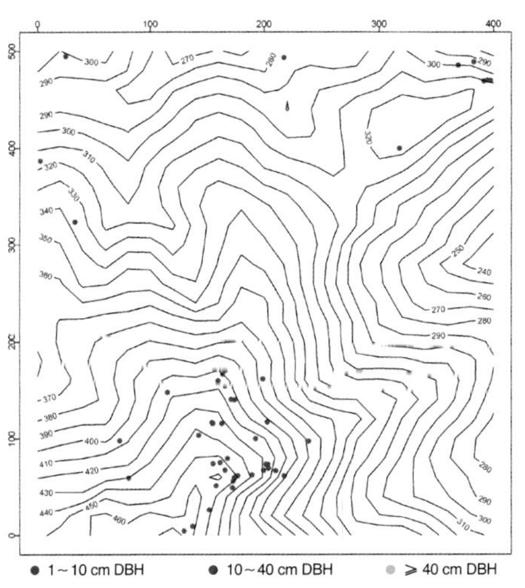

个体分布图 Distribution of individuals

径级分布表 DBH class

胸径等级 (Diameter class) (cm)	个体数 (No. of individuals in the plot)	比例 (Proportion) (%)
1～2	23	47.92
2～5	17	35.42
5～10	8	16.67
10～20	0	0.00
20～30	0	0.00
30～60	0	0.00
≥60	0	0.00

19 粗壮润楠（两广润楠） cūzhuàngrùnnán | Robust Machilus

Machilus robusta W.W.Smith
樟科 | Lauraceae

代码（SpCode）= MACROB
个体数（Individual number/20 hm²）= 48
最大胸径（Max DBH）= 34.3 cm
重要值排序（Importance value rank）= 74

常绿乔木，高15～20m，树皮粗糙，黑灰色。枝条粗壮，圆柱形，具皮孔。叶狭椭圆状卵形至倒卵状椭圆形或近长圆形，厚革质，两面极无毛，中脉叶背十分凸起，变红色。花序生于枝顶和先端叶腋，多花，分枝。果球形，径2.5～3cm，熟时蓝黑色，果梗粗壮深红色。花期1～4月，果期4～6月。

Evergreen trees, to 15-20 m tall. Bark blackish gray, rough. Branchlets thick, terete, lenticellate. Leaf blade sometimes slightly glaucous abaxially, elliptic to obovate-elliptic or suboblong, thickly leathery, glabrous on both surfaces, midrib raised abaxially, reddish when fresh, concave adaxially. Panicles terminal or subterminal, much branched, many flowered. Fruit globose, 2.5-3 cm in diam., blue-black when mature, fruiting pedicel thickened. Fl. Jan.-Apr., fr. Apr.-Jun..

野外识别特征：

1. 树皮粗糙，黑灰色；
2. 枝粗壮，圆柱形，具皮孔；
3. 叶较大，长10～20(26)cm，宽（2.5）5.5～8.5cm。
4. 叶中脉在叶背很凸起，略带红色。

Key notes for identification:

1. Bark blackish gray, rough;
2. Branchlets thick, terete, lenticellate;
3. Leaf blade slightly larger, length 10-20 (26) cm, breadth (2.5) 5.5-8.5 cm;
4. Leaves midrib raised abaxially, reddish.

叶 Leaf
摄影：吴林芳 Photo by: Wu Linfang

叶背 Leaf abaxial surface
摄影：吴林芳 Photo by: Wu Linfang

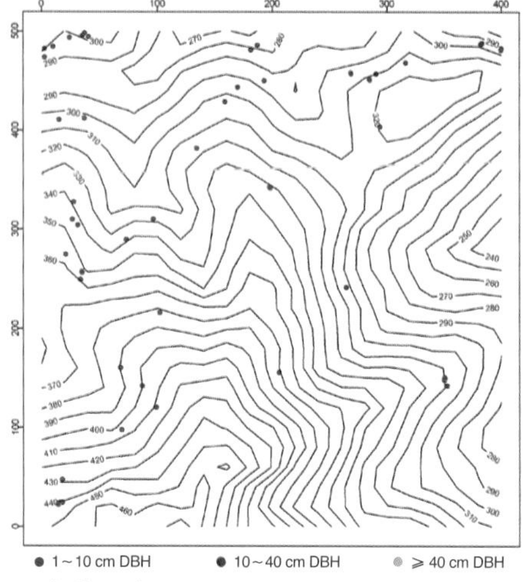

个体分布图 Distribution of individuals

径级分布表 DBH class

胸径等级 (Diameter class) (cm)	个体数 (No. of individuals in the plot)	比例 (Proportion) (%)
1～2	10	20.83
2～5	8	16.67
5～10	23	47.92
10～20	7	14.58
20～30	0	0.00
30～60	0	0.00
≥60	0	0.00

20 绒楠

róngnán | Woolly Machilus

Machilus velutina Champ. ex Benth.
樟科 | Lauraceae

代码（SpCode）= MACVEL
个体数（Individual number/20 hm^2）= 236
最大胸径（Max DBH）= 36.3 cm
重要值排序（Importance value rank）= 48

常绿乔木，高达18m，胸径40cm。枝、芽、叶背及花序均密被锈色茸毛。叶狭倒卵形、椭圆形或狭卵形，革质，中脉叶面稍凹，叶背很凸起。花序顶生，近无梗，花黄色，被锈色茸毛。果球形，径约4mm，紫红色。花期10～12月，果期翌年2～3月。

Evergreen trees, to 18 m tall, 40 cm d.b.h.. Branch, bud, leaf abaxial surface and inflorescence densely ferruginous tomentose. Leaf blade narrowly obovate, elliptic, or narrowly ovate, leathery, midrib abaxially raised, slightly concave adaxially. Inflorescences terminal, peduncle very short, flowers yellowish, densely ferruginous tomentose. Fruit purplish red, globose, ca. 4 mm in diam.. Fl. Oct.-Dec., fr. Feb.-Mar. of next year.

野外识别特征：

1. 各部密被锈色茸毛，别于黄绒润楠的黄褐色茸毛；
2. 叶基楔形，别于黄绒润楠的多少圆形；
3. 叶先端渐狭或短渐尖。

Key notes for identification:

1. All parts densely ferruginous tomentose, as *M. grijsii* densely yellow-brown velutinous;
2. Leaves base cuneate, as *M. grijsii* base ± rounded;
3. Leaves apex attenuate or shortly acuminate.

果及叶背 Fruit and leaf abaxial surface
摄影：吴林芳 Photo by: Wu Linfang

花枝 Flowering branch
摄影：吴林芳 Photo by: Wu Linfang

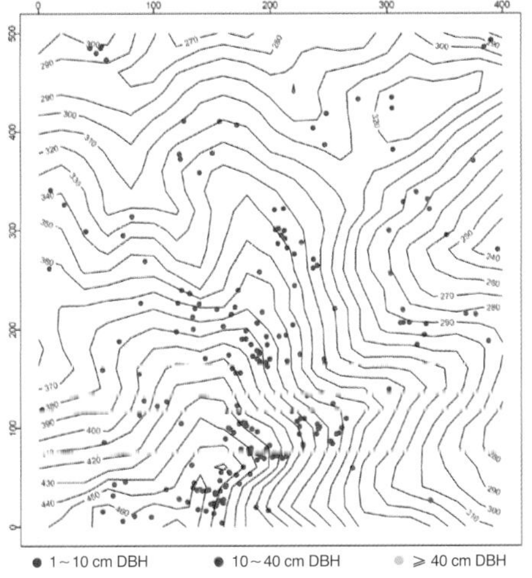

个体分布图 Distribution of individuals

径级分布表 DBH class

胸径等级 (Diameter class) (cm)	个体数 (No. of individuals in the plot)	比例 (Proportion) (%)
1～2	114	48.31
2～5	88	37.29
5～10	25	10.59
10～20	4	1.69
20～30	3	1.27
30～60	2	0.85
≥60	0	0.00

21 新木姜子

xīnmùjiāngzǐ | Aurata Neolitsea

Neolitsea aurata (Hay.) Koidz.

樟科 | Lauraceae

代码（SpCode）= NEOAUR

个体数（Individual number/20 hm^2）= 4

最大胸径（Max DBH）= 2.2 cm

重要值排序（Importance value rank）= 153

常绿乔木，高达14m。树皮灰褐色，幼枝和叶柄被锈色或黄褐色短柔毛或无毛。叶互生或聚生枝顶呈轮生状，长圆形、椭圆形至长圆状披针形，革质，叶面绿色无毛，叶背密被金黄色绢毛；离基三出脉，侧脉每边3～4条。伞形花序3～5个簇生枝顶或节间。果椭圆柱形。花期2～3月，果期9～10月。

Evergreen trees, up to 14 m tall. Bark brown, young branchlets and petioles ferruginous or yellow-brown pubescent or glabrous. Leaves alternate or clustered toward apex of branchlet, leaf blade oblong, elliptic, oblanceolate, densely golden yellow sericeous abaxially glabrous adaxially, leathery, triplinerved, lateral veins 3 or 4 pairs. Umbels 3-5-fascicled toward apex of branchlet or internode. Fruit ellipsoid. Fl. Feb.-Mar., fr. Sep.-Oct..

野外识别特征：

1. 幼枝和叶柄被锈色或黄褐色短柔毛；
2. 叶背密被金黄色绢毛；
3. 离基三出脉。

Key notes for identification:

1. Young branchlets and petioles ferruginous or yellow-brown pubescent;
2. Leaf blade densely golden yellow sericeous abaxially;
3. Leaf blade triplinerved.

花 Flower

摄影：叶育石 Photo by: Ye Yushi

叶 Leaf

摄影：叶育石 Photo by: Ye Yushi

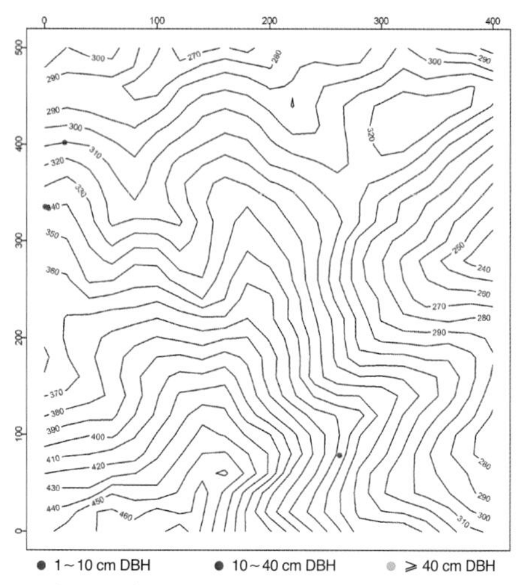

个体分布图 Distribution of individuals

径级分布表 DBH class

胸径等级 (Diameter class) (cm)	个体数 (No. of individuals in the plot)	比例 (Proportion) (%)
1～2	3	75.00
2～5	1	25.00
5～10	0	0.00
10～20	0	0.00
20～30	0	0.00
30～60	0	0.00
≥60	0	0.00

22 锈叶新木姜子（柬埔寨新木姜） xiùyèxīnmùjiāngzǐ | Cambodia Neolitsea

Neolitsea cambodiana Lec.
樟科 | Lauraceae

代码（SpCode）= NEOCAM
个体数（Individual number/20 hm²）= 273
最大胸径（Max DBH）= 11.6 cm
重要值排序（Importance value rank）= 49

常绿乔木，高8～12m。幼时密被锈色茸毛。叶3～5片近轮生，椭圆状披针形，革质，幼时两面密被锈色茸毛，后渐落，羽状脉或近离基三出脉。伞形花序5～7个簇生，无梗或近无梗，每花序有花4～5朵。果球形。花期10～12月，果期翌年7～8月。

Evergreen trees, 8-12 m tall. Young branchlets densely ferruginous tomentose or yellow-brown appressed pubescent. Leaves 3-5-subverticillate; leaf blade oblong-lanceolate, oblong-elliptic, leathery, densely ferruginous tomentose on both surfaces when young and becoming glabrous, pinninerved or subtriplinerved. Umbels 5-7-clustered, sessile or subsessile, 4- or 5-flowered. Fruit globose. Fl. Oct.-Dec., fr. Jul.-Aug. of next year.

野外识别特征：

1. 小枝轮生或近轮生，幼时密被锈色茸毛；
2. 叶3～5片轮生，幼时密被锈色茸毛后渐落；
3. 羽状脉或近离基三出脉。

Key notes for identification:

1. Branchlets verticillate or subverticillate denselyferruginous tomentose;
2. Leaves 3-5-subverticillate densely ferruginous tomentose on both surfaces when young and becoming glabrous;
3. Leaf blade pinninerved or subtriplinerved.

叶 Leaf
摄影：叶育石 Photo by: Ye Yushi

果 Fruit
摄影：叶育石 Photo by: Ye Yushi

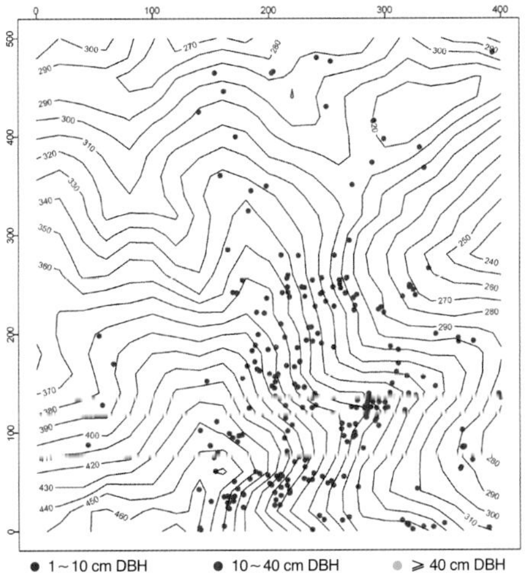

个体分布图 Distribution of individuals

径级分布表 DBH class

胸径等级 (Diameter class) (cm)	个体数 (No. of individuals in the plot)	比例 (Proportion) (%)
1～2	103	37.73
2～5	139	50.92
5～10	28	10.26
10～20	3	1.10
20～30	0	0.00
30～60	0	0.00
≥60	0	0.00

23 鸭公树

yāgōngshù | Chu's Neolitsea

Neolitsea chuii Merr.

樟科 | Lauraceae

代码（SpCode）= NEOCHU

个体数（Individual number/20 hm^2）= 7

最大胸径（Max DBH）= 3.1 cm

重要值排序（Importance value rank）= 121

常绿乔木，高8～18m。树皮灰青色或灰褐色。除花序外其他各部均无毛。叶互生或聚生枝顶呈轮生状，椭圆形至长圆状椭圆形，革质，离基三出脉。伞形花序腋生或侧生，多个密集。果椭圆体或近球形，花期9～10月，果期12月。

Evergreen trees, 8-18m tall. Bark gray-cyan or gray-brown. Glabrous throughout except inflorescences. Leaves alternate or crowded toward apex of the branchlets, leaf blade elliptic to oblong-elliptic or ovate-elliptic, triplinerved, leathery. Umbels axillary or lateral, clustered. Fruit ellipsoid or subglobose. Fl. Sep.-Oct., fr. -Dec..

野外识别特征：

1. 叶离基三出脉,常聚生枝顶；
2. 叶背粉绿色，常被白粉；
3. 全株除花序外其他各部均无毛。

Key notes for identification:

1. Leaves triplinerved, usually clustered toward apex of branchlet;
2. Leaf blade palegreen abaxially, usually with whiting;
3. Glabrous throughout except inflorescences.

叶 Leaf

摄影：吴林芳 Photo by: Wu Linfang

花枝 Flowering branch

摄影：叶育石 Photo by: Ye Yushi

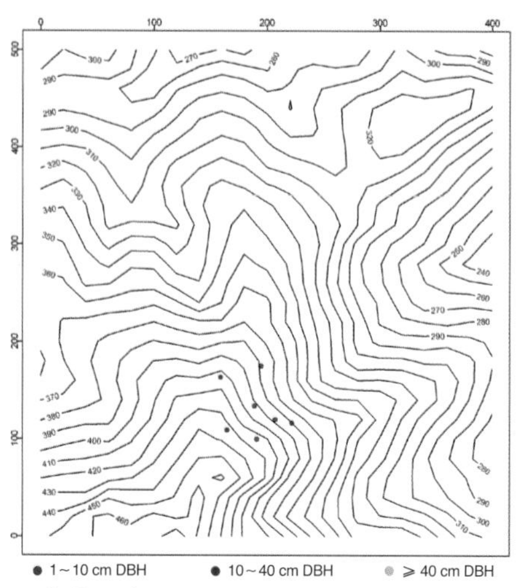

个体分布图 Distribution of individuals

径级分布表 DBH class

胸径等级 (Diameter class) (cm)	个体数 (No. of individuals in the plot)	比例 (Proportion) (%)
1～2	6	85.71
2～5	1	14.29
5～10	0	0.00
10～20	0	0.00
20～30	0	0.00
30～60	0	0.00
≥60	0	0.00

24 灰白新木姜子（小新木姜） huībáixīnmùjiāngzǐ | Pale Neolitsea

Neolitsea pallens (D. Don) Momiyama et Hara
樟科 | Lauraceae

代码（SpCode）= NEOPAL
个体数（Individual number/20 hm^2）= 1327
最大胸径（Max DBH）= 27.2 cm
重要值排序（Importance value rank）= 20

常绿乔木，高10～15m。小枝纤细，幼时被淡黄褐色短柔毛。叶互生或3～5片聚生枝顶，椭圆形或椭圆状披针形，长3～8cm，宽1～4cm，薄革质，离基三出脉，叶面绿色有光泽，叶背灰白。伞形花序生叶腋。果椭圆体至球形，长8mm，径6mm，花期3～5月，果期6～7月。

Evergreen trees, 10-15 m tall. Branchlets slender, young branchlets yellowish brown pubescent and becoming glabrous. Leaves alternate or 3-5 clustered toward apex of branchlet, leaf blade elliptic or elliptic-lanceolate, 1-4 cm × 3-8 cm, thinly leathery, leaf blade green glabrate adaxially, pale white abaxially, triplinerved. Umbels axillary. Fruit ellipsoid-globose, 8 mm long, 6 mm in diam. Fl. Mar.-May, fr. Jun.-Jul..

野外识别特征：

1. 叶离基三出脉，常聚生枝顶，纸质；
2. 叶一般较小，长3～8cm，宽1～4cm；
3. 叶面亮绿，叶背灰白。

Key notes for identification:

1. Leaves aggregate, chartaceous, triplinerved;
2. Leaf blade usually small, length 3-8 cm, breadth 1-4 cm;
3. Leaf blade green glabrate adaxially, pale white abaxially.

叶 Leaf
摄影：吴林芳 Photo by: Wu Linfang

叶背 Leaf abaxial surface
摄影：吴林芳 Photo by: Wu Linfang

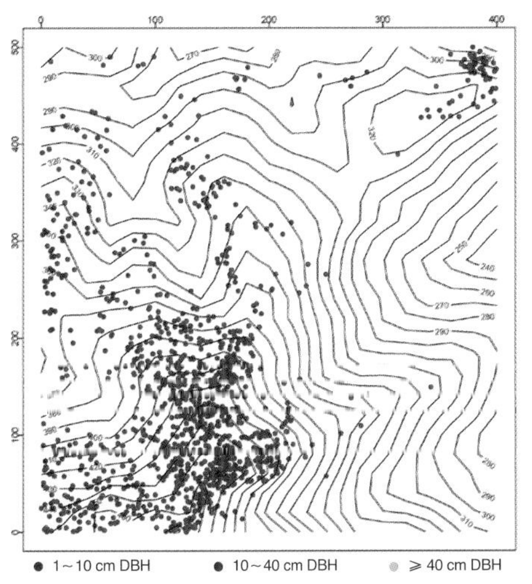

个体分布图 Distribution of individuals

径级分布表 DBH class

胸径等级 (Diameter class) (cm)	个体数 (No. of individuals in the plot)	比例 (Proportion) (%)
1～2	414	31.20
2～5	505	38.06
5～10	295	22.23
10～20	111	8.36
20～30	2	0.15
30～60	0	0.00
≥60	0	0.00

25 黄叶树

huángyèshù | Hainan Xanthophyllum

Xanthophyllum hainanense Hu
远志科 | Polygalaceae

代码（SpCode）= NEOCHU
个体数（Individual number/20 hm^2）= 1852
最大胸径（Max DBH）= 54.1 cm
重要值排序（Importance value rank）= 14

常绿乔木，高达20m。树皮暗灰色，具细纵裂。小枝纤细，圆柱形，无毛。叶片革质，卵状椭圆形至长圆状披针形，两面无毛，叶脉两面凸起。总状花序或小型圆锥花序腋生或顶生，总花梗和花梗密被短柔毛，花白黄色。核果球形，淡黄色，径1.5～2cm，被柔毛。花期3～5月，果期4～7月。

Evergreen trees, up to 20 m tall. Bark dark gray, longitudinally thinly fissured. Branchlets terete, slender, glabrous. Leaf blade ovate-elliptic to oblong-lanceolate, leathery, both surfaces glabrous, midvein and lateral veins raised on both surfaces. Racemes or small panicles axillary or terminal, petals yellow-white, peduncles and pedicels densely pubescent. Drupe globose, yellowish, 1.5-2 cm in diam.. Fl. Mar.-May, fr. Apr.-Jul..

野外识别特征：

1. 小枝纤细，黄绿色，无毛，略呈之字形；
2. 叶两面无毛，脉两面凸起；
3. 叶先端长尾尖。

Key notes for identification:

1. Branchlets yellowgreen, slender, glabrous, branchlets zigzag;
2. Leaf blade both surfaces glabrous, veins raised on both surfaces;
3. Leaf blade apex long acuminate.

叶 Leaf
摄影：吴林芳 Photo by: Wu Linfang

叶背 Leaf abaxial surface
摄影：吴林芳 Photo by: Wu Linfang

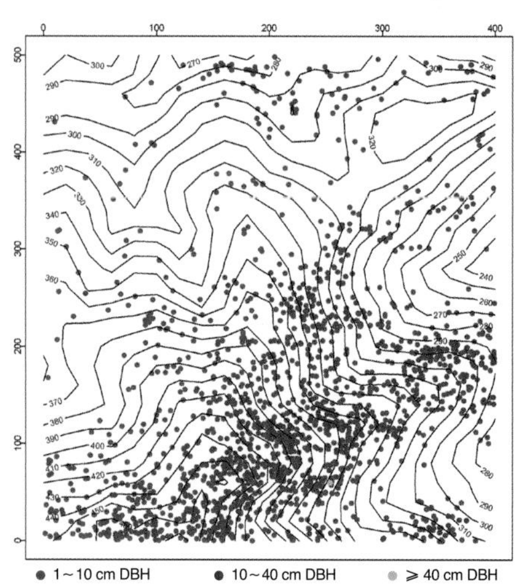

个体分布图 Distribution of individuals

径级分布表 DBH class

胸径等级 (Diameter class) (cm)	个体数 (No. of individuals in the plot)	比例 (Proportion) (%)
1～2	436	23.53
2～5	854	46.09
5～10	378	20.40
10～20	151	8.15
20～30	26	1.40
30～60	7	0.38
≥60	0	0.00

26 土沉香（白木香） tǔchénxiāng I Incense Tree

Aquilaria sinensis (Lour.) Spreng.
瑞香科 I Thymelaeaceae

代码（SpCode）= AQUSIN
个体数（Individual number/20 hm^2）= 25
最大胸径（Max DBH）= 29.5 cm
重要值排序（Importance value rank）= 88

常绿乔木，高5～15m。树皮暗灰色，平滑。叶革质，圆形、椭圆形、长圆形有时倒卵形，先端锐尖或急尖而具短尖头，两面无毛，小脉纤细近平行，不明显，边缘有时被稀疏的柔毛。花序顶生，夜晚具芳香，黄绿色，多朵组成伞形花序。蒴果卵球形，种子褐色卵球形，基部附着体长，约1.5cm×0.4cm。花期春夏，果期夏秋。

Evergreen trees, 5-15 m tall. Bark dark gray, smooth. Leaf blade orbicular or elliptic to oblong, sometimes obovate, leathery, both surfaces glabrous, apex acuminate or acute, apiculate, veins and veinlets slender, subparallel, obscure. Inflorescence terminal, many flowered. Flowers fragrant at night, yellowish green. Capsule void; seeds dark brown, ovoid, funicle conspicuous, ca. 1.5 cm × 0.4 cm, longer than seed. Fl. spring-summer, fr. summer-autumn.

种子 Seeds
摄影：吴林芳 Photo by: Wu Linfang

果 Fruit
摄影：吴林芳 Photo by: Wu Linfang

花 Flower
摄影：吴林芳 Photo by: Wu Linfang

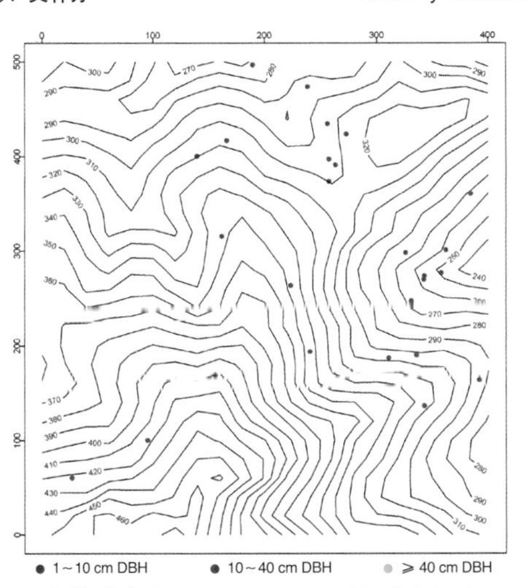

个体分布图 Distribution of individuals

径级分布表 DBH class

胸径等级 (Diameter class) (cm)	个体数 (No. of individuals in the plot)	比例 (Proportion) (%)
1～2	10	40.00
2～5	10	40.00
5～10	2	8.00
10～20	2	8.00
20～30	1	4.00
30～60	0	0.00
≥60	0	0.00

27 了哥王

liǎogēwáng | Indian Wikstroemia

Wikstroemia indica (L.) C. A. Mey.
瑞香科 | Thymelaeaceae

代码（SpCode）= WIKIND
个体数（Individual number/ 20 hm^2）= 5
最大胸径（Max DBH）= 1.7 cm
重要值排序（Importance value rank）= 130

常绿灌木，高2m或更高。小枝红褐色，无毛。叶对生，纸质至近革质，倒卵形、椭圆状长圆形或披针形，干时棕红色，无毛。花黄绿色，数朵组成顶生头状花序，花序梗短，不超过5mm，无毛。核果椭圆体，长7～8mm，熟时红色至暗紫色。花果期夏秋间。

Evergreen shrubs 2 m or more tall. Branches reddish brown, glabrous. Leaves opposite, leaf blade reddish brown on both surfaces when dried, obovate, elliptic-oblong, or lanceolate, papery to subleathery, both surfaces glabrous, Inflorescences terminal, capitate, several flowered; peduncle short then 5 mm, glabrous. Drupe red to dark purple, ellipsoid, 7-8 mm long. Fl. and fr. summer-autumn.

野外识别特征：

1. 花序梗粗短，不超过5mm；
2. 叶倒卵形、椭圆状长圆形或披针形。

Key notes for identification:

1. Peduncle shorter than 5 mm, slightly stout;
2. Leaf blade obovate, elliptic-oblong, or lanceolate.

花枝 Flowering branch
摄影：吴林芳 Photo by: Wu Linfang

果枝 Fruiting branch
摄影：吴林芳 Photo by: Wu Linfang

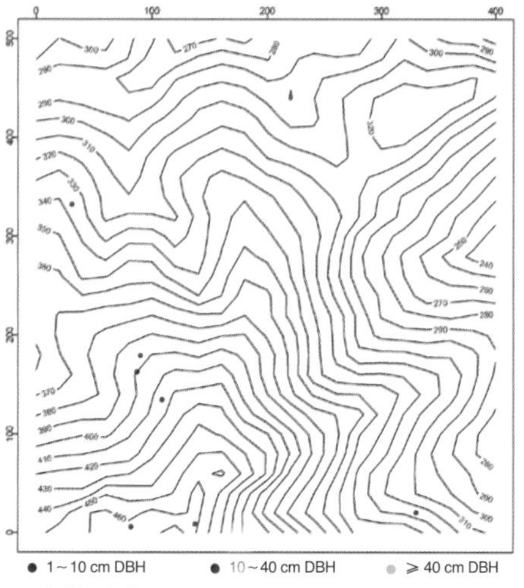

个体分布图 Distribution of individuals

径级分布表 DBH class

胸径等级 (Diameter class) (cm)	个体数 (No. of individuals in the plot)	比例 (Proportion) (%)
1～2	5	100.0
2～5	0	0.00
5～10	0	0.00
10～20	0	0.00
20～30	0	0.00
30～60	0	0.00
≥60	0	0.00

28 细轴荛花

xìzhóuráohuā | Nodding Wikstroemia

Wikstroemia nutans Champ. ex Benth.
瑞香科 | Thymelaeaceae

代码（SpCode）= WIKNUT
个体数（Individual number/20 hm^2）= 18
最大胸径（Max DBH）= 2.5 cm
重要值排序（Importance value rank）= 96

常绿灌木，高1～2m或更高。小枝圆柱形，红褐色，无毛。叶对生，膜质至纸质，卵形、卵状椭圆形至卵状披针形，两面无毛。花萼黄绿色，3～8朵组成顶生近头状的总状花序，花序梗细长，0.7～3cm。果椭圆形，长约7mm，熟时深红色。花期春季至初夏，果期夏季至秋季。

Evergrren shrubs 1-2 m or more tall. Branches reddish or grayish brown, terete, glabrous. Leaves opposite, leaf blade ovate or ovate-elliptic to ovate-lanceolate, membranous to papery, both surfaces glabrous. Inflorescences terminal, shortly racemose, 3-8-flowered, peduncle 0.7-3 cm, slender, glabrous, calyx yellowish green. Drupe dark red, ellipsoid, ca. 7 mm long. Fl. spring-early summer, fr. summer-autumn.

野外识别特征：

1. 花梗细长，约7～30mm；
2. 叶卵形、卵状椭圆形至卵状披针形。

Key notes for identification:

1. Peduncle 7-30 mm, slender;
2. Leaf blade ovate or ovate-elliptic to ovate-lanceolate.

果枝 Fruiting branch
摄影：吴林芳 Photo by: Wu Linfang

花枝 Flowering branch
摄影：吴林芳 Photo by: Wu Linfang

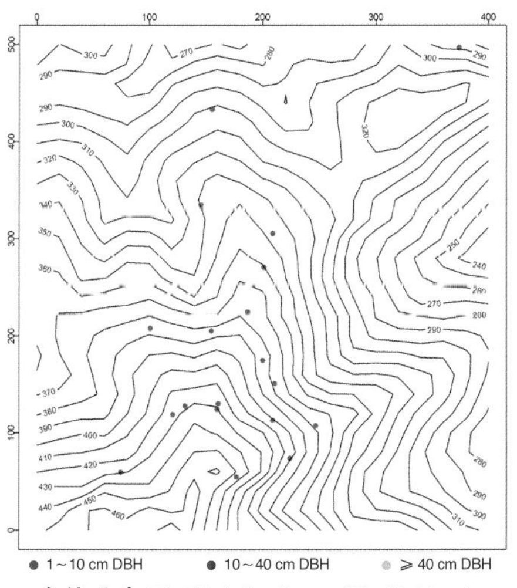

个体分布图 Distribution of individuals

径级分布表 DBH class

胸径等级 (Diameter class) (cm)	个体数 (No. of individuals in the plot)	比例 (Proportion) (%)
1～2	12	66.67
2～5	6	33.33
5～10	0	0.00
10～20	0	0.00
20～30	0	0.00
30～60	0	0.00
≥60	0	0.00

29 小果山龙眼（越南山龙眼） xiǎoguǒshānlóngyǎn | Cochinchina Helicia

Helicia cochinchinensis Lour.

山龙眼科 | Proteaceae

代码（SpCode）= HELCOC

个体数（Individual number/20 hm^2）= 5

最大胸径（Max DBH）= 6.5 cm

重要值排序（Importance value rank）= 139

常绿乔木或灌木，高4～20m。枝叶均无毛。叶薄革质或纸质，长椭圆形或披针形，幼树叶叶缘具疏齿，大树叶常全缘或上半部具疏齿。总状花序腋生，长7～14（20）cm，无毛；花被白色或淡黄色。果椭圆形，长1～1.5cm；果皮薄革质，厚不超过0.5mm，黑色。花期6～10月，果期11月至翌年3月。

Evergreen trees or shrubs, 4-20 m tall. Branchlets and leaves glabrous. Leaf blade elliptic, obovate-oblong or lanceolate, papery or thinly leathery, remotely serrate when young, margin entire when old or serrate on apical half. Inflorescences axillary, 7-14 (-20) cm, glabrescent, perianth whitish or yellowish. Fruit ellipsoid, 1-1.5 cm, pericarp less than 0.5 mm thick, black, thinly leathery. Fl. Jun.-Oct., fr. Nov.-Mar. of next year.

花枝 Flowering branch

摄影：吴林芳 Photo by: Wu Linfang

果枝 Fruiting branch

摄影：叶育石 Photo by: Ye Yushi

幼叶 Young leave

摄影：吴林芳 Photo by: Wu Linfang

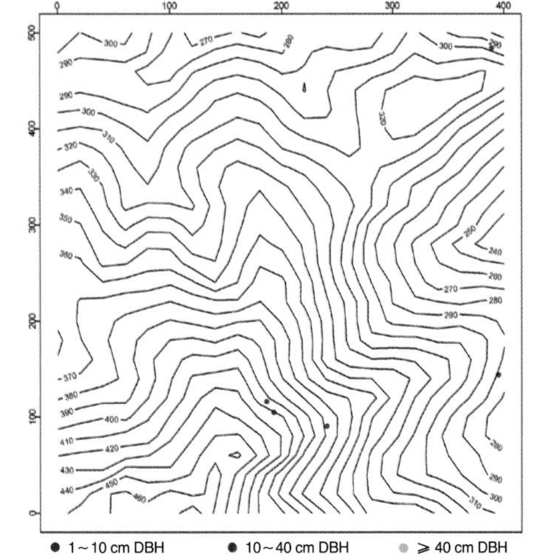

个体分布图 Distribution of individuals

径级分布表 DBH class

胸径等级 (Diameter class) (cm)	个体数 (No. of individuals in the plot)	比例 (Proportion) (%)
1～2	0	0.00
2～5	1	20.00
5～10	4	80.00
10～20	0	0.00
20～30	0	0.00
30～60	0	0.00
≥60	0	0.00

30 网脉山龙眼 wǎngmàishānlóngyǎn I Reticula Helicia

Helicia reticulata W. T. Wang
山龙眼科 I Proteaceae

代码（SpCode）= PHELRET
个体数（Individual number/20 hm^2）= 68
最大胸径（Max DBH）= 19.2 cm
重要值排序（Importance value rank）= 65

常绿灌木或乔木，高3～10m。嫩叶、芽被褐色或锈色短毛，小枝和成熟叶则无毛。叶革质，长圆形至倒披针形，边缘具疏齿或细齿，网脉在两面均凸起明显。总状花序腋生或生于小枝已脱落的腋部，无毛。果椭圆体状，长1.5～1.8cm；果皮厚革质，约1mm，黑色。花期5～7月，果期10～12月。

Evergreen shrubs or trees, 3-10 m tall. New leaf and bud have brown or ferruginous pubescent, branchlets and mature leaves glabrescent. Leaf blade oblong, obovate, or oblanceolate, leathery, midvein and secondary veins raised on both surfaces. Inflorescences axillary or ramiflorous, glabrous. Fruit black, ellipsoid, 1.5-1.8 cm long, pericarp ca. 1 mm thick, thickly leathery, black. Fl. May-Jul., fr. Oct.-Dec..

果 Fruit
摄影：吴林芳 Photo by: Wu Linfang

叶 Leaf
摄影：吴林芳 Photo by: Wu Linfang

花 Flower
摄影：吴林芳 Photo by: Wu Linfang

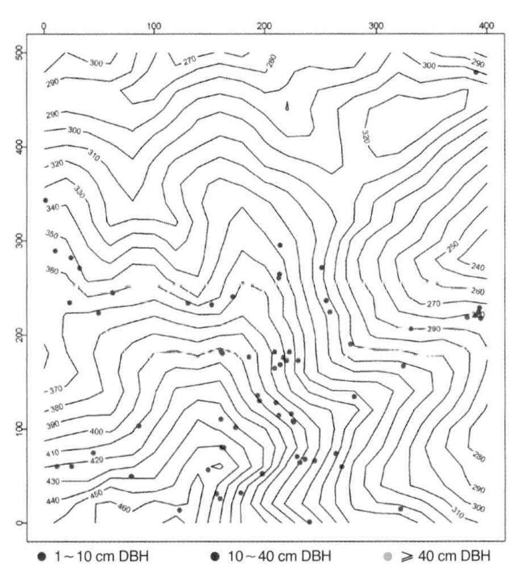

个体分布图 Distribution of individuals

径级分布表 DBH class

胸径等级 (Diameter class) (cm)	个体数 (No. of individuals in the plot)	比例 (Proportion) (%)
1～2	6	8.82
2～5	19	27.94
5～10	37	54.41
10～20	6	8.82
20～30	0	0.00
30～60	0	0.00
≥60	0	0.00

31 光叶海桐

guāngyèhǎitóng | Glabrous Pittosporum

Pittosporum glabratum Lindl.
海桐花科 | Pittosporaceae

代码（SpCode）= PITGLA
个体数（Individual number/20 hm^2）= 3
最大胸径（Max DBH）= 1.4 cm
重要值排序（Importance value rank）= 142

常绿灌木，高2～3m。嫩枝无毛，老枝有皮孔。叶聚生枝顶，薄革质，狭矩圆形或倒披针形，边缘平整，有时稍皱折。伞形花序1～4枝簇生于枝顶叶腋，多花。蒴果椭圆形，有时长筒形，3片裂开；种子近圆形，红色，花期3～8月，果期6～12月。

Evergreen shrubs, 2-3 m tall. Young branchlets glabrous, old branchlets lenticellate. Leaves clustered at branchlet apex, thinly leathery, leaf blade obovate to narrowly so, oblong, oblanceolate, margin flat, sometimes slightly rugose. Inflorescences 1-4 in leaf axils at branchlet apex, umbellate or corymbose, many flowered. Capsule ellipsoid, sometimes long tubelike, dehiscing by 3 valves, seeds red, subglobose. Fl. Mar.-Aug., fr. Jun.-Dec..

花枝 Flowering branch
摄影：吴林芳 Photo by: Wu Linfang

果枝 Fruiting branch
摄影：吴林芳 Photo by: Wu Linfang

叶背 Leaf abaxial surface
摄影：吴林芳 Photo by: Wu Linfang

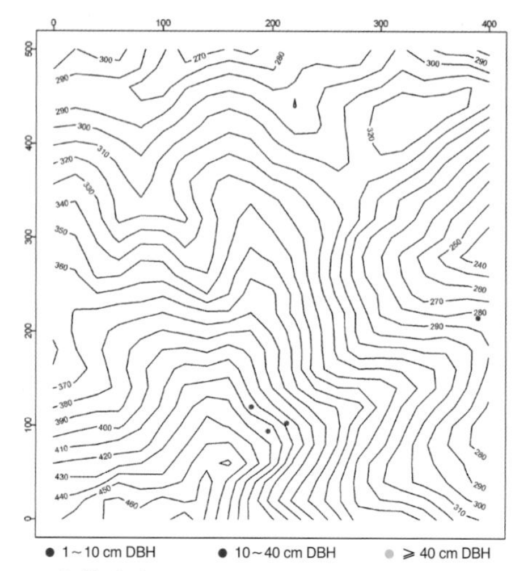

个体分布图 Distribution of individuals

径级分布表 DBH class

胸径等级 (Diameter class) (cm)	个体数 (No. of individuals in the plot)	比例 (Proportion) (%)
1～2	3	100.0
2～5	0	0.00
5～10	0	0.00
10～20	0	0.00
20～30	0	0.00
30～60	0	0.00
≥60	0	0.00

32 球花脚骨脆（嘉赐树） qíuhuājiǎogǔcuì | Casearia

Casearia glomerata Roxb.
天料木科 | Samydaceae

代码（SpCode）= CASGLO
个体数（Individual number/20 hm^2）= 5
最大胸径（Max DBH）= 12.0 cm
重要值排序（Importance value rank）= 135

常绿乔木或灌木，高4～10m。树皮灰褐色。幼枝有棱和柔毛，老枝无毛。叶薄革质，排成二列，长椭圆形至卵状椭圆形。花黄绿色，10～15朵或更多组成团伞花序，腋生。蒴果卵形，通常不裂，种子卵形多数，花期8～12月，果期10月至翌年春季。

Evergreen trees or shrubs, 4-10 m tall. Bark grey brown. Young branchlets angulate, pubescent. Leaf distichous, elliptic, lanceolate, or oblong, thinly leathery. Flowers 10-15 or more in axillary glomerules, yellowish. Capsule ellipsoid usually not dehiscence, seeds several, ovoid. Fl. Aug.-Dec., fr. Oct.-spring of next year.

野外识别特征：
1. 叶互生，排成二列；
2. 叶两面无毛；
3. 树皮光滑，具明显皮孔。

Key notes for identification:
1. Leaves alternate, distichous;
2. Leaf blade glabrous on both surfaces;
3. Bark smoothly, conspicuously lenticellate.

叶 Leaf
摄影：吴林芳 Photo by: Wu Linfang

叶背及果枝 Leaf abaxial surface & fruiting branch
摄影：吴林芳 Photo by: Wu Linfang

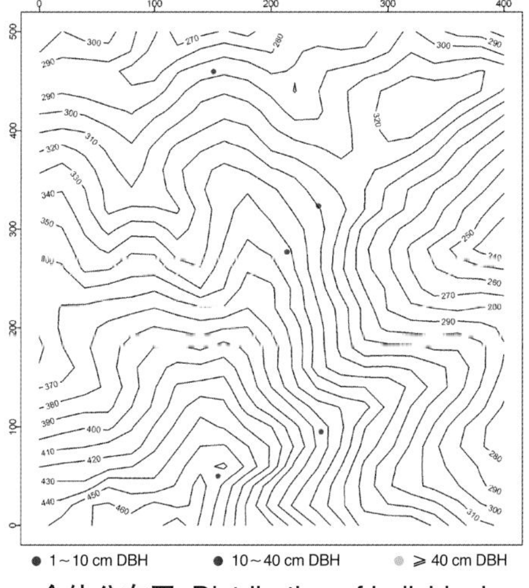

个体分布图 Distribution of individuals

径级分布表 DBH class

胸径等级 (Diameter class) (cm)	个体数 (No. of individuals in the plot)	比例 (Proportion) (%)
1～2	3	60.00
2～5	1	20.00
5～10	0	0.00
10～20	1	20.00
20～30	0	0.00
30～60	0	0.00
≥60	0	0.00

33 毛叶脚骨脆（毛叶嘉赐树） máoyèjiǎogǔcuì | Villous Casearia

Casearia velutina Bl.

天料木科 | Samydaceae

代码（SpCode）= CASVEL

个体数（Individual number/20 hm^2）= 38

最大胸径（Max DBH）= 4.8 cm

重要值排序（Importance value rank）= 79

常绿小乔木，高可达10m。枝呈之字形，密被锈色茸毛。叶柄粗壮，密被褐色短毛；叶椭圆形至长椭圆形，厚纸质，叶背密被黄褐色长茸毛。花少数至多数，组成团伞形花序。果实长椭圆形，长1~1.2cm，无毛，肉质，熟时黄色，种子卵形多数，花期3~5月，果期4~6月。

Evergreen small trees, to 10 m tall. Branchlets zigzag, densely rusty pubescent. Petioles stout, densely brown-pubescent, leaf blade elliptic or elliptic-oblong, thickly papery, densely yellowish brown villose abaxially. Flowers several to many in glomerules. Capsule ellipsoid, 1-1.2 cm, glabrous, fleshy, yellow when mature, seeds many, ovoid. Fl. Mar.-May, fr. Apr.-Jun..

野外识别特征：

1. 叶互生，排成二列；
2. 叶背密被黄褐色长茸毛；
3. 树皮光滑，具明显皮孔。

Key notes for identification:

1. Leaves alternate, distichous;
2. Leaf blade densely yellowish brown villose abaxially;
3. Bark smoothly, conspicuously lenticellate.

花枝 Flowering branch

摄影：叶育石 Photo by: Ye Yushi

叶背 Leaf abaxial surface

摄影：吴林芳 Photo by: Wu Linfang

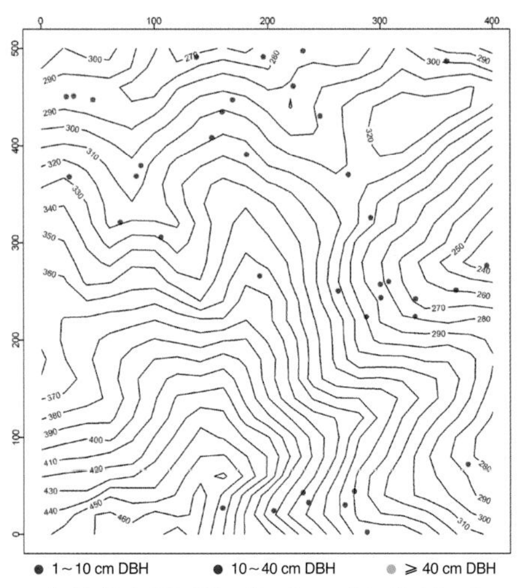

个体分布图 Distribution of individuals

径级分布表 DBH class

胸径等级 (Diameter class) (cm)	个体数 (No. of individuals in the plot)	比例 (Proportion) (%)
1~2	19	50.00
2~5	19	50.00
5~10	0	0.00
10~20	0	0.00
20~30	0	0.00
30~60	0	0.00
≥60	0	0.00

34 天料木

tiānliàomù | Cochinchina Homalium

Homalium cochinchinense (Lour.) Druce
天料木科 | Samydaceae

代码（SpCode）= HOMCOC
个体数（Individual number/20 hm^2）= 45
最大胸径（Max DBH）= 13.0 cm
重要值排序（Importance value rank）= 80

落叶小乔木或灌木，高2～10m。小枝圆柱形，幼时密被带黄色短柔毛，老枝无毛，有明显纵棱。叶纸质，宽椭圆状长圆形至倒卵状长圆形。花多数，单个或簇生排成总状，有时略分枝，被黄色短柔毛。蒴果倒圆锥状，近无毛。花全年，果9～12月。

Deciduous shrubs or small trees, 2-10 m tall. Branchlets terete, densely yellowish pubescent when young, gradually glabrescent. Leaf blade broadly elliptic, elliptic-oblong, or obovate, papery. Inflorescence racemelike, rachis pubescent, hairs spreading. Capsule inverted cone, subglabrous. Fl. all year, fr. Sep.-Dec..

野外识别特征：

1. 嫩枝密被带黄色短柔毛，老枝无毛；
2. 叶宽椭圆状长圆形至倒卵状长圆形，纸质，叶柄极短，叶缘具齿；
3. 花排成总状花序，被毛。

Key notes for identification:

1. Twigs densely yellowish pubescent gradually glabrescent;
2. Leaf blade broadly elliptic, elliptic-oblong, or obovate, thickly papery, margin obtusely serrate, petiole extremely short;
3. Inflorescence racemelike, rachis pubescent.

花序 Inflorescence
摄影：吴林芳 Photo by: Wu Linfang

叶 Leaf
摄影：吴林芳 Photo by: Wu Linfang

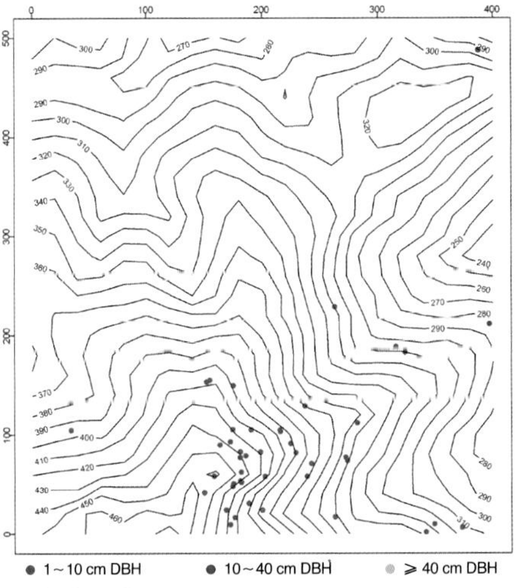

个体分布图 Distribution of individuals

径级分布表 DBH class

胸径等级 (Diameter class) (cm)	个体数 (No. of individuals in the plot)	比例 (Proportion) (%)
1～2	7	15.56
2～5	21	46.67
5～10	15	33.33
10～20	2	4.44
20～30	0	0.00
30～60	0	0.00
≥60	0	0.00

35 杨桐 yángtóng | Millet Adinandra

Adinandra millettii (Hooker et Arnott) Benth.
山茶科 | Theaceae

代码（SpCode）= ADIMIL
个体数（Individual number/20 hm^2）= 3
最大胸径（Max DBH）= 10.1 cm
重要值排序（Importance value rank）= 162

常绿灌木或小乔木，高2～10（16）m。嫩枝及顶芽被毛。单叶互生，2列，革质，长圆状椭圆形，全缘，少有在上半部略有细牙齿。花单朵腋生，白色，子房3室，花柱单一不分叉，宿存。果圆球形，直径约1cm，熟时黑色，种子多数。花期5～7月，果期8～10月。

Evergreen shrubs or small trees, 2-10 (16) m tall. Young branchlets and terminal buds appressed pubescent. Leaves alternate, distichous, leaf blade oblong-elliptic, leathery, margin entire or apically sparsely serrate. Flowers axillary, solitary, petals white, ovary globose, 3-loculed, style completely united. Fruit black when mature, globose, ca. 1 cm in diam., many seeded. Fl. May-Jul., fr. Aug.-Oct..

叶背 Leaf abaxial surface
摄影：吴林芳 Photo by: Wu Linfang

花枝 Flowering branch
摄影：吴林芳 Photo by: Wu Linfang

果枝 Fruiting branch
摄影：吴林芳 Photo by: Wu Linfang

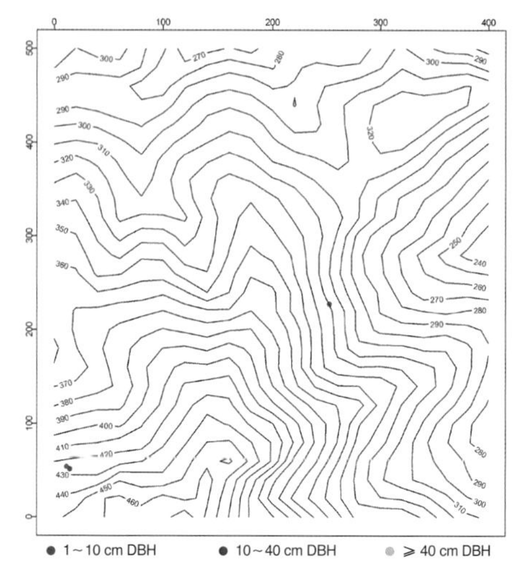

个体分布图 Distribution of individuals

径级分布表 DBH class

胸径等级 (Diameter class) (cm)	个体数 (No. of individuals in the plot)	比例 (Proportion) (%)
1～2	0	0.00
2～5	1	33.33
5～10	1	33.33
10～20	1	33.33
20～30	0	0.00
30～60	0	0.00
≥60	0	0.00

36 岗柃

gānglíng | Groff Eurya

Eurya groffii Merr.
山茶科 | Theaceae

代码（SpCode）= EURGRO
个体数（Individual number/20 hm^2）= 1
最大胸径（Max DBH）= 4.3 cm
重要值排序（Importance value rank）= 177

常绿灌木或小乔木，高2～7（～10）m。嫩枝圆柱形，密被黄褐色披散柔毛，小枝圆形近无毛。叶披针形，边缘有锯齿，叶面无毛，叶背被长毛。花白色，单性，1～9簇生叶腋，花药不分格，子房3室，花柱3裂。浆果圆球形，径4mm，黑色。花期9～11月，果期翌年4～6月。

Evergreen shrubs or small trees, 2-7 (-10) m tall. Current year branchlets terete, densely yellowish brown spreading villous, young branches pubescent or glabrescent. Leaf blade lanceolate, adaxially glabrous, abaxially appressed pilose. Flowers white, unisexual, axillary, solitary or to 9 in a cluster, anthers not locellate, ovary 3-loculed, style 3-parted. Fruit purplish black 4 mm in diam.. Fl. Sep.-Nov., fr. Apr.-Jun next year.

野外识别特征：

1. 嫩枝圆柱形，密被黄褐色披散柔毛；
2. 叶披针形至长披针形，边缘具细齿；
3. 花柱稍长，长2～2.5mm。

Key notes for identification:

1. Twigs terete, densely yellowish brown spreading villous;
2. Leaf blade lanceolate to oblong-lanceolate, margin closely serrulate;
3. Style relatively long, length 2-2.5 mm.

花枝 Flowering branch
摄影：吴林芳 Photo by: Wu Linfang

果枝 Fruiting branch
摄影：吴林芳 Photo by: Wu Linfang

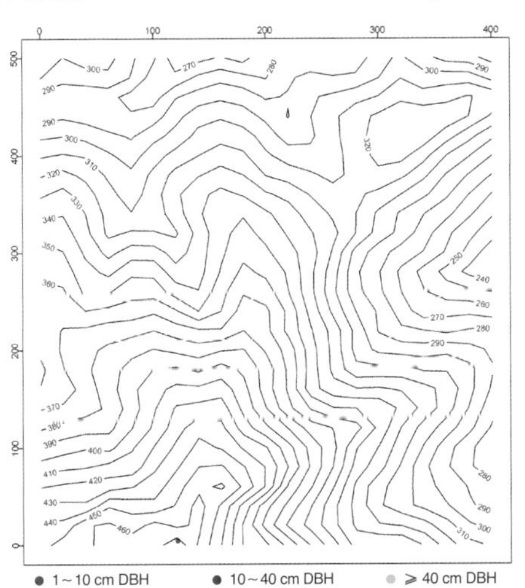

● 1～10 cm DBH ● 10～40 cm DBH ● ≥ 40 cm DBH
个体分布图 Distribution of individuals

径级分布表 DBH class

胸径等级 (Diameter class) (cm)	个体数 (No. of individuals in the plot)	比例 (Proportion) (%)
1～2	0	0.00
2～5	1	100.00
5～10	0	0.00
10～20	0	0.00
20～30	0	0.00
30～60	0	0.00
≥60	0	0.00

37 黑柃

hēilíng | Black Eurya

Eurya macartneyi Champ.
山茶科 | Theaceae

代码（SpCode）= EURMAC
个体数（Individual number/20 hm^2）= 553
最大胸径（Max DBH）= 29.0 cm
重要值排序（Importance value rank）= 32

常绿灌木或小乔木，高2～7m。树皮黑褐色。嫩枝粗壮圆柱形，无毛，顶芽披针形，无毛。叶革质，长圆状椭圆形至椭圆形，长6～14cm，宽2～4.5cm，两面无毛。花1～4朵簇生叶腋，花药不具分格，子房3室，花柱3，离生。浆果圆球形，径5mm，熟时紫黑色。花期11月至翌年1月，果期翌年6～8月。

Evergreen shrubs or small trees, 2-7 m tall. Bark blackish brown. Current year branchlets stout, terete, glabrous, terminal buds lanceolate, glabrous. Leaf blade oblong-elliptic to elliptic, 6-14 cm × 2-4.5 cm, leathery, both surfaces glabrous. Flowers axillary, solitary or to 4 in a cluster, anthers not locellate, ovary 3-loculed, styles 3, distinct. Fruit purplish black 5 mm in diam.. Fl. Nov.-Jan. of next year, fr. Jun.-Aug. of next year.

花枝 Flowering branch
摄影：吴林芳 Photo by: Wu Linfang

果 Fruit
摄影：吴林芳 Photo by: Wu Linfang

叶 Leaf
摄影：吴林芳 Photo by: Wu Linfang

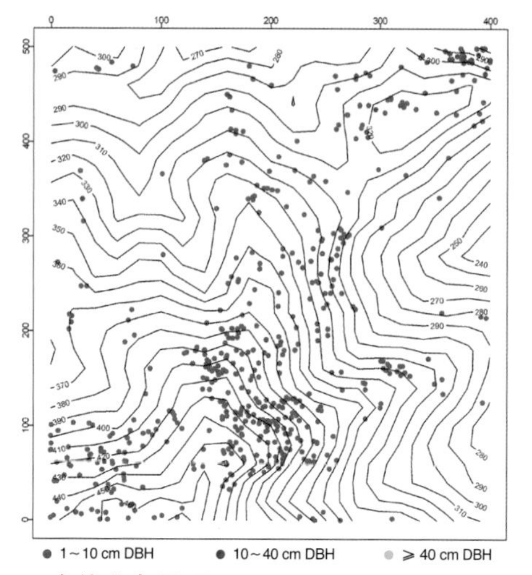

个体分布图 Distribution of individuals

径级分布表 DBH class

胸径等级 (Diameter class) (cm)	个体数 (No. of individuals in the plot)	比例 (Proportion) (%)
1～2	19	3.44
2～5	218	39.42
5～10	305	55.15
10～20	10	1.81
20～30	1	0.18
30～60	0	0.00
≥60	0	0.00

38 细齿叶柃 xìchǐyèlíng | Shining Eurya

Eurya nitida Korth.
山茶科 | Theaceae

代码（SpCode）= EURNIT
个体数（Individual number/20 hm²）= 1
最大胸径（Max DBH）= 10.3 cm
重要值排序（Importance value rank）= 174

常绿灌木或小乔木，高2～5m。全株无毛，嫩枝稍纤细，2棱。叶薄革质，椭圆形至倒卵状长圆形，长4～6cm，宽1.5～2.5cm。花1～4朵簇生叶腋，花药不具分格，花柱细长，顶端3浅裂。浆果圆球形，径3～4mm。花期11～翌年1月，果期翌年7～9月。

Evergreen shrubs or small trees, 2-5 m tall. Whole plant glabrous, current year branchlets slender, 2-ribbed. Leaf blade elliptic, oblong-elliptic, or obovate-oblong, 4-6 cm × 1.5-2.5 cm, thinly leathery. Flowers axillary, solitary or to 4 in a cluster, anthers not locellate, style slender, apically 3-lobed. Fruit globose, 3-4 mm in diam.. Fl. Nov.-Jan. of next year, fr. Jul.-Sep. of next year.

野外识别特征：

1. 叶与米碎花相似，但全株无毛；
2. 嫩枝稍纤细，2棱，无毛；
3. 叶薄革质或革质，椭圆形至倒卵状椭圆形。

Key notes for identification:

1. Leaves very similar to *E. chiensis*, but whole plant glabrous;
2. Current year branchlets slender, 2-ribbed, glabrous;
3. Leaf blade elliptic oblong-elliptic, or obovate-oblong, thinly leathery.

果枝 Fruiting branch
摄影：叶育石 Photo by: Ye Yushi

叶 Leaf
摄影：卓书斌 Photo by: Zhuo Shubin

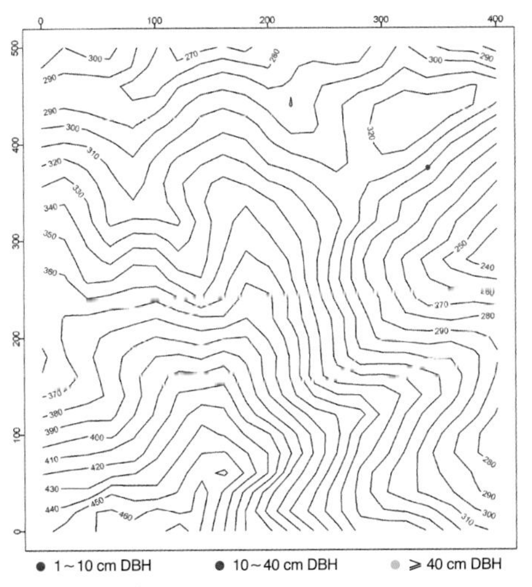

个体分布图 Distribution of individuals

径级分布表 DBH class

胸径等级 (Diameter class) (cm)	个体数 (No. of individuals in the plot)	比例 (Proportion) (%)
1～2	0	0.00
2～5	0	0.00
5～10	0	0.00
10～20	1	100.00
20～30	0	0.00
30～60	0	0.00
≥60	0	0.00

39 木荷

mùhé | Gugertree

Schima superba Gardn. et Champ.
山茶科 | Theaceae

代码（SpCode）= SCHSUP
个体数（Individual number/20 hm²）= 2290
最大胸径（Max DBH）= 89.0 cm
重要值排序（Importance value rank）= 2

常绿大乔木，高25m。嫩枝通常无毛。叶革质或薄革质，椭圆形至长圆状椭圆形，叶背无毛，边缘有钝齿。花生于枝顶叶腋，常多朵排列成总状花序，花白色，径2～3cm，萼片半圆形。蒴果近球形，直径1.5～2cm。花期6～8月，果期10～12月。

Evergreen trees, to 25 m tall. Young branches glabrous or puberulent. Leaf blade elliptic to oblong-elliptic, thinly leathery to leathery, abaxially glabrous, margin undulately. Flowers in a raceme, 2-3 cm in diam, sepals semiorbicular, petals white. Capsule subglobose, 1-2 cm in diam.. Fl. Jun.-Aug., fr. Oct.-Dec..

树干 Trunk
摄影：吴林芳 Photo by: Wu Linfang

果枝 Fruiting branch
摄影：吴林芳 Photo by: Wu Linfang

花枝 Flowering branch
摄影：曹洪麟 Photo by: Cao Honglin

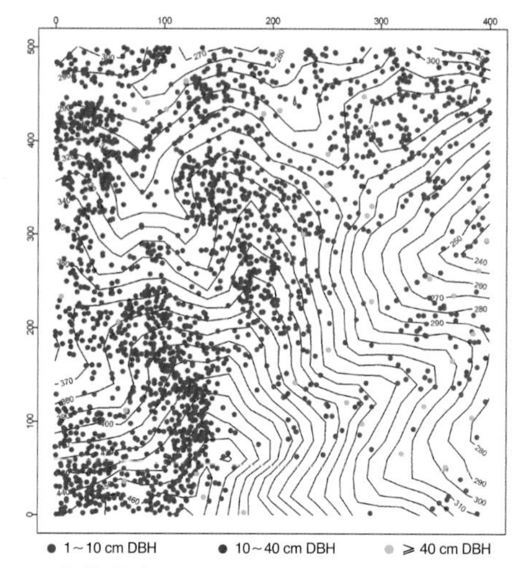

个体分布图 Distribution of individuals

径级分布表 DBH class

胸径等级 (Diameter class) (cm)	个体数 (No. of individuals in the plot)	比例 (Proportion) (%)
1～2	6	0.26
2～5	18	0.79
5～10	136	5.94
10～20	1183	51.66
20～30	770	33.62
30～60	174	7.60
≥60	3	0.13

40 厚皮香

hòupíxiāng I Nakedanthe Ternstroemia

Ternstroemia gymnanthera (Wight et Arn.) Bedd.
山茶科 I Theaceae

代码（SpCode）= TERGYM
个体数（Individual number/20 hm^2）= 10
最大胸径（Max DBH）= 10.4 cm
重要值排序（Importance value rank）= 114

常绿灌木或乔木，高1.5～10（～15）m。全株无毛。叶革质，椭圆形、椭圆状倒卵形至长圆状倒卵形，全缘或稀上部有疏齿。花通常生于当年生无叶的小枝上或生于叶腋。浆果圆球形或扁球形，径1～1.5cm，肉质假种皮熟时红色。花期5～7月，果期9～11月。

Evergreen shrubs or trees, 1.5-10 (-15) m tall. Whole plant glabrous. Leaf blade obovate, oblong-obovate, or broadly elliptic, leathery, margin entire or apically sparsely serrate. Flowers axillary, solitary or several clustered on leafless branchlets. Fruit globose or oblate, red when mature, 1-1.5 cm in diam.. Fl. May-Jul., fr. Sep.-Nov..

树干 Trunk
摄影：吴林芳 Photo by: Wu Linfang

花枝 Flowering branch
摄影：吴林芳 Photo by: Wu Linfang

果枝 Fruiting branch
摄影：吴林芳 Photo by: Wu Linfang

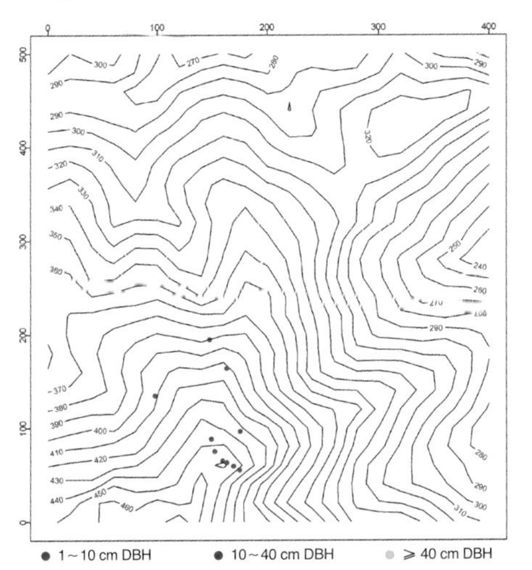

个体分布图 Distribution of individuals

径级分布表 DBH class

胸径等级 (Diameter class) (cm)	个体数 (No. of individuals in the plot)	比例 (Proportion) (%)
1～2	0	0.00
2～5	1	10.00
5～10	6	60.00
10～20	3	30.00
20～30	0	0.00
30～60	0	0.00
≥60	0	0.00

41 五列木

wǔlièmù | Common Pentaphylax

Pentaphylax euryoides Gardn. et Champ.
五列木科 | Pentaphylacaceae

代码（SpCode）= PENEUR
个体数（Individual number/20 hm^2）= 1
最大胸径（Max DBH）= 1.5 cm
重要值排序（Importance value rank）= 189

常绿灌木或乔木，高4～10 m。小枝圆柱形，灰褐色，无毛。单叶互生，革质，卵形或卵状长圆形或长圆状披针形，全缘略反卷，无毛；叶面具槽。总状花序腋生或顶生，花白色。蒴果椭圆状，褐黑色，种子红棕色，先端极压扁或呈翅状。花期4～6月，果期10～11月。

Evergreen shrubs or trees, 4-10 m tall. Branchlets grayish brown, terete, glabrous. Leaf blade alternate, ovate, oblong-ovate, or oblong-lanceolate, leathery, margin entire and slightly revolute, both surfaces glabrous, petiole adaxially grooved. Inflorescence raceme, axillary or subterminal, petals white. Capsule purplish dark brown, ellipsoid, seeds reddish brown, apex very compressed or winged. Fl. Apr.-Jun., fr. Oct.-Nov..

果枝 Fruiting branch
摄影：吴林芳 Photo by: Wu Linfang

花序 Inflorescence
摄影：吴林芳 Photo by: Wu Linfang

叶 Leaf
摄影：吴林芳 Photo by: Wu Linfang

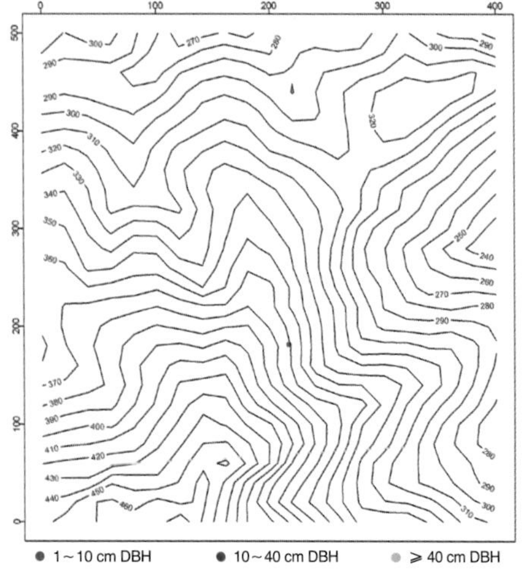

个体分布图 Distribution of individuals

径级分布表 DBH class

胸径等级 (Diameter class) (cm)	个体数 (No. of individuals in the plot)	比例 (Proportion) (%)
1～2	1	100.00
2～5	0	0.00
5～10	0	0.00
10～20	0	0.00
20～30	0	0.00
30～60	0	0.00
≥60	0	0.00

42 桃金娘

táojīnniáng | Rose Myrtle

Rhodomyrtus tomentosa (Ait.) Hassk.
桃金娘科 | Myrtaceae

代码（SpCode）= RHOTOM
个体数（Individual number/20 hm^2）= 1
最大胸径（Max DBH）= 2.1 cm
重要值排序（Importance value rank）= 180

常绿灌木。嫩枝被灰白色柔毛。叶对生，革质，叶片椭圆形或倒卵形，离基三出脉，直达先端且相结合。花有长梗，常单生，紫红色，直径2～4cm。浆果卵状壶形，长1.5～2cm，宽1～1.5cm，熟时紫黑色。花期4～5月，果期夏末秋初。

Evergreen shrubs. Current year branchlets grayish tomentose. Leaves opposite, leaf blade elliptic to obovate, leathery, triplinerved, meeting at apex. Flowers long-pedicellate, often solitary, purplish red, 2-4 cm in diam.. Berry purplish black when mature, urceolate, 1.5-2 cm × 1-1.5 cm. Fl. Apr.-May, fr. Late summer to early autumn.

花枝 Flowering branch
摄影：吴林芳 Photo by: Wu Linfang

叶 Leaf
摄影：吴林芳 Photo by: Wu Linfang

果枝 Fruiting branch
摄影：吴林芳 Photo by: Wu Linfang

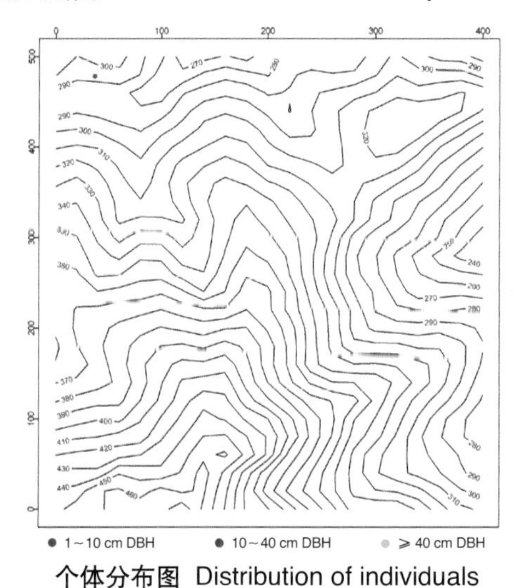

个体分布图 Distribution of individuals

径级分布表 DBH class

胸径等级 (Diameter class) (cm)	个体数 (No. of individuals in the plot)	比例 (Proportion) (%)
1～2	0	0.00
2～5	1	100.00
5～10	0	0.00
10～20	0	0.00
20～30	0	0.00
30～60	0	0.00
≥60	0	0.00

43 肖蒲桃

xiāopútáo | Willow-leaf Acmena

Syzygium acuminatissimum (Blume) DC.
桃金娘科 | Myrtaceae

代码（SpCode）= SYZACU
个体数（Individual number/20 hm^2）= 1526
最大胸径（Max DBH）= 69.7 cm
重要值排序（Importance value rank）= 12

常绿乔木，高20m。小枝圆形或有钝棱。叶革质，对生，有油腺点，卵状披针形或狭披针形，先端尾状渐尖。聚伞花序排列成圆锥花序，顶生，花序轴有棱，花3朵聚生，两性。果球形，径1.5cm，熟时紫黑色。花期7～10月。

Evergreen trees, up to 20 m tall. Branchlets terete or obtusely angulate. Leaves opposite, with oil glands, leaf blade leathery, ovate-lanceolate to narrowly lanceolate, apex caudate-acuminate. Inflorescence terminal, 3 flowered cymes arranged into panicles, peduncle ridged, bisexual. Fruit drupaceous, blackish purple when mature, ca. 1.5 cm in diam.. Fl. Jul.-Oct..

花枝 Flowering branch
摄影：吴林芳 Photo by: Wu Linfang

叶背 Leaf abaxial surface
摄影：吴林芳 Photo by: Wu Linfang

果枝 Fruiting branch
摄影：吴林芳 Photo by: Wu Linfang

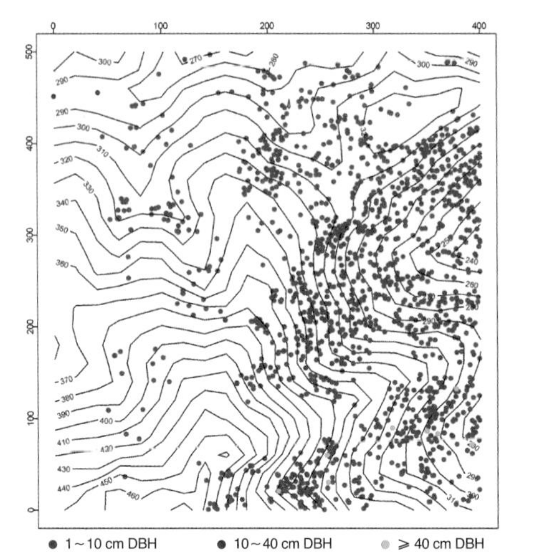

个体分布图 Distribution of individuals

径级分布表 DBH class

胸径等级 (Diameter class) (cm)	个体数 (No. of individuals in the plot)	比例 (Proportion) (%)
1～2	162	10.62
2～5	394	25.82
5～10	369	24.18
10～20	413	27.06
20～30	136	8.91
30～60	51	3.34
≥60	1	0.07

44 赤楠

chìnán | Box-leaved Syzygium

Syzygium buxifolium Hook. et Arn.
桃金娘科 | Myrtaceae

代码（SpCode）= SYZBUX
个体数（Individual number/20 hm^2）= 2
最大胸径（Max DBH）= 6.4 cm
重要值排序（Importance value rank）= 172

常绿灌木或小乔木。嫩枝有棱。叶对生，革质，较小，阔椭圆形至椭圆形，长1.5～3cm，宽1～2cm。聚伞花序顶生，长约1cm，有花数朵。果实球形，直径5～7mm，熟时红到紫黑色。花期6～8月，果期10～12月。

Evergreen shrubs or small trees. Current year branchlets angled. Leaves opposite, leaf blade broadly elliptic to elliptic, 1.5-3 cm × 1-2 cm, leathery. Inflorescences terminal, cymes, ca. 1 cm, several-flowered. Fruit red turning purplish black, globose, 5-7 mm in diam.. Fl. Jun.-Aug., fr. Oct.-Dec..

枝叶 Branch and leaves
摄影：吴林芳 Photo by: Wu Linfang

果枝 Fruiting branch
摄影：董安强 Photo by: Dong Anqiang

花枝 Flowering branch
摄影：吴林芳 Photo by: Wu Linfang

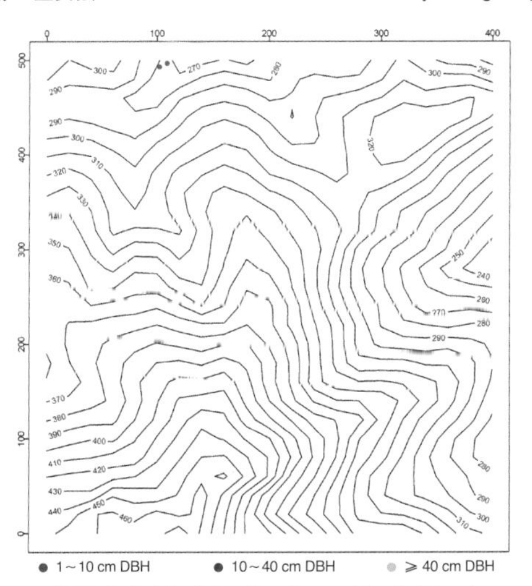

个体分布图 Distribution of individuals

径级分布表 DBH class

胸径等级 (Diameter class) (cm)	个体数 (No. of individuals in the plot)	比例 (Proportion) (%)
1～2	0	0.00
2～5	1	50.0
5～10	1	50.0
10～20	0	0.00
20～30	0	0.00
30～60	0	0.00
≥60	0	0.00

45 子凌蒲桃 zǐlíngpútáo | Champion's Syzygium

Syzygium championii (Benth.) Merr. et Perry
桃金娘科 | Myrtaceae

代码（SpCode）= SYZCHA
个体数（Individual number/20 hm^2）= 401
最大胸径（Max DBH）= 31.9 cm
重要值排序（Importance value rank）= 35

常绿灌木至乔木。嫩枝4棱，干后灰白色。叶革质，狭长圆形至椭圆形，长3～6（～9）cm，宽1～2cm。聚伞花序顶生，有时腋生，有花6～10朵，花蕾棒状，长1cm。果实长椭圆形，长12mm，红色。花期8～11月，果期10～12月。

Evergreen shrubs to trees. Young branchlets 4-angled, grayish white when dry. Leaf blade leathery, narrowly oblong to elliptis, 3-6 (-9) cm × 1-2 cm. Cymes terminal or sometimes axillary, 6-10-flowered, flower-buds clavate, ca. 1 cm. Fruit red, long-ellipsoidal, ca. 1.2 cm. Fl. Aug.-Nov., fr. Oct.-Dec..

野外识别特征：

1. 树皮灰褐色，光滑；
2. 嫩枝4棱，枝红色；
3. 叶狭长圆形至椭圆形，先端急尖，带不及1cm的尖头；
4. 花蕾棒状，果长椭圆形，红色。

Key notes for identification:

1. Bark gray-brown, smoothly;
2. Young branchlets 4-angled, branchlets red;
3. Leaf blade narrowly oblong to elliptis, apex acute and usually with a cusp less than 1 cm;
4. Flower buds clavate, fruit red, long ellipsoid.

叶 Leaf
摄影：吴林芳 Photo by: Wu Linfang

叶背 Leaf abaxial surface
摄影：吴林芳 Photo by: Wu Linfang

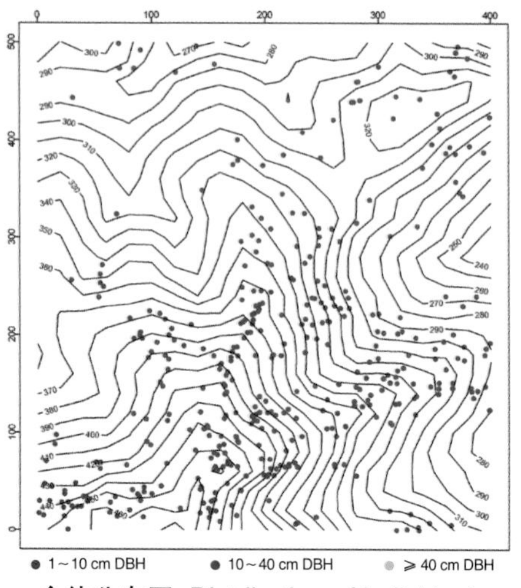

个体分布图 Distribution of individuals

径级分布表 DBH class

胸径等级 (Diameter class) (cm)	个体数 (No. of individuals in the plot)	比例 (Proportion) (%)
1～2	133	33.17
2～5	156	38.90
5～10	55	13.72
10～20	41	10.22
20～30	14	3.49
30～60	2	0.50
≥60	0	0.00

46 红鳞蒲桃（小花蒲桃） hónglínpútáo | Hance's Syzygium

Syzygium hancei Merr. et Perry
桃金娘科 | Myrtaceae

代码（SpCode）= SYZHAN
个体数（Individual number/20 hm²）= 3
最大胸径（Max DBH）= 23.8 cm
重要值排序（Importance value rank）= 144

常绿灌木至乔木，高达20m。嫩枝圆形，红褐色。叶片革质，边脉离边缘约0.5mm，叶柄短，3～6mm。圆锥花序腋生，花小，1～1.5cm；花蕾倒卵形，约2mm，花瓣分离。果球形，径5～8mm。花期7～9月，果期11月至翌年1月。

Evergreen shrubs or trees, to 20 m tall. Young branchlets terete, red-brown. Leaf blade leathery, intramarginal veins ca. 0.5 mm from margin, petiole short, 3-6 mm. Inflorescences axillary, paniculate cymes, 1-1.5 cm, many-flowered, flower buds obovoid, ca. 2 mm, petals distinct. Fruit globose, 5-8 mm in diam.. Fl. Jul.-Sep., fr. Nov.-Jan. of next year.

野外识别特征：
1. 树皮红褐色，粗糙；
2. 嫩枝圆形，红褐色；
3. 叶狭椭圆形、长圆形或倒卵形，先端钝或略尖；
4. 花蕾倒卵形，果球形，红至黑。

Key notes for identification:
1. Bark red-brown, rough;
2. Twigs terete, red-brown;
3. Leaf blade narrowly elliptic, oblong, or obovate, apex obtuse to slightly acute;
4. Flower buds obovoid, fruit red to black, globose.

果序 Infructescence
摄影：吴林芳 Photo by: Wu Linfang

叶 Leaf
摄影：吴林芳 Photo by: Wu Linfang

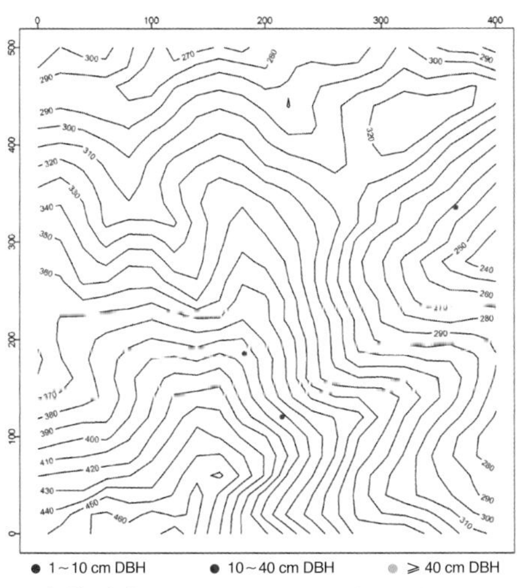

个体分布图 Distribution of individuals

径级分布表 DBH class

胸径等级 (Diameter class) (cm)	个体数 (No. of individuals in the plot)	比例 (Proportion) (%)
1～2	1	33.33
2～5	1	33.33
5～10	0	0.00
10～20	0	0.00
20～30	1	33.33
30～60	0	0.00
≥60	0	0.00

47 蒲桃 pútáo | Rose Apple

Syzygium jambos (L.) Alston
桃金娘科 | Myrtaceae

代码（SpCode）= SYZJAM
个体数（Individual number/20 hm²）= 22
最大胸径（Max DBH）= 18.8 cm
重要值排序（Importance value rank）= 100

常绿乔木，高10m。小枝圆形。叶厚革质，披针形或长圆形，长8～26cm，宽2～4.5cm，侧脉间隔大，约7～15mm，叶柄长5～10mm。聚伞花序顶生，数朵，花大，白色，径3～4cm，花瓣分离。果球形，径2.5～5cm，果皮肉质，熟时黄色。花期3～4月，果期5～6月或11～12月。

Evergreen trees, to 10 m tall. Branchlets terete. Petiole 5-10 mm, leaf blade lanceolate, ovate-lanceolate, 8-26 cm × 2-4.5 cm, thickly leathery, secondary veins 7-15 mm apart. Inflorescences usually terminal cymes with several flowers, flowers white, 3-4 cm in diam., petals distinct. Fruit yellow when mature, globose, 2.5-5 cm in diam., pericarp fleshy. Fl. Mar.-Apr., fr. May-Jun. or Nov.-Dec..

果 Fruit
摄影：吴林芳 Photo by: Wu Linfang

花 Flower
摄影：吴林芳 Photo by: Wu Linfang

叶 Leaf
摄影：吴林芳 Photo by: Wu Linfang

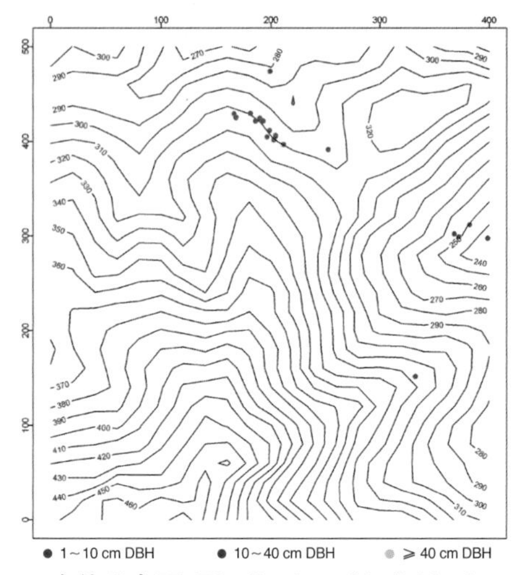

个体分布图 Distribution of individuals

径级分布表 DBH class

胸径等级 (Diameter class) (cm)	个体数 (No. of individuals in the plot)	比例 (Proportion) (%)
1～2	0	0.00
2～5	8	36.36
5～10	7	31.82
10～20	7	31.82
20～30	0	0.00
30～60	0	0.00
≥60	0	0.00

48 广东蒲桃 guǎngdōngpútáo I Kwangtung Syzygium

Syzygium kwangtungense (Merr.) Merr. et Perry
桃金娘科 I Myrtaceae

代码（SpCode）= SYZKWA
个体数（Individual number/20 hm^2）= 3
最大胸径（Max DBH）= 12.1 cm
重要值排序（Importance value rank）= 164

常绿小乔木，高5m。嫩枝圆形或稍压扁，老枝褐色。叶片革质，椭圆形至狭椭圆形，叶柄3～5mm，边脉离边缘约1mm。圆锥花序顶生或近顶生，花短小，常3朵簇生，花瓣连合成帽状。果球形，径7～9mm。花期6～7月，果期10～12月。

Evergreen small trees, to 5 m tall. Current year branchlets terete or slightly compressed, old branches brown. Petiole 3-5 mm, leaf blade elliptic to narrowly elliptic, leathery, intramarginal veins ca. 1 mm from margin. Inflorescences terminal or axillary, paniculate cymes, coherent. Fruit, globose, 7-9 mm in diam.. Fl. Jun.-Jul., fr. Oct.-Dec..

野外识别特征：

1. 树皮褐色，略光滑；
2. 嫩枝圆形或稍压扁，褐色；
3. 叶椭圆形至狭椭圆形，先端钝或略尖；
4. 花蕾倒圆锥形；果球形，黑色。

Key notes for identification:

1. Bark brown, slightly smoothly;
2. Twigs terete or slightly compressed brown;
3. Leaf blade elliptic to narrowly elliptic, apex obtuse to slightly acute;
4. Flower buds obconic, fruit black, globose.

果枝 Fruiting branch
摄影：易绮斐 Photo by: Yi Qifei

叶 Leaf
摄影：叶育石 Photo by: Ye Yushi

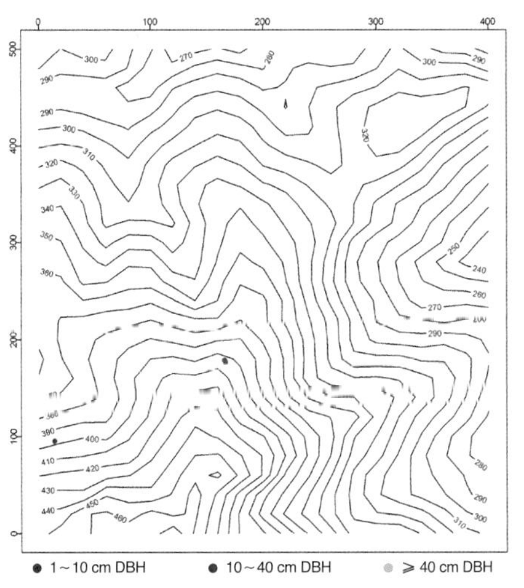

个体分布图 Distribution of individuals

径级分布表 DBH class

胸径等级 (Diameter class) (cm)	个体数 (No. of individuals in the plot)	比例 (Proportion) (%)
1～2	0	0.00
2～5	1	33.33
5～10	1	33.33
10～20	1	33.33
20～30	0	0.00
30～60	0	0.00
≥60	0	0.00

49 山蒲桃（白车） shānpútáo | Levine's Syzygium

Syzygium levinei (Merr.) Merr. et Perry
桃金娘科 | Myrtaceae

代码（SpCode）= SYZLEV
个体数（Individual number/20 hm^2）= 407
最大胸径（Max DBH）= 61.0 cm
重要值排序（Importance value rank）= 33

常绿乔木，高达25m。嫩枝圆形，有糠秕，灰白或灰色。叶革质，椭圆形至卵状椭圆形，叶柄中等长，5～7mm。圆锥花序顶生或上部腋生，花蕾倒卵形，长4～5mm，花瓣分离。果近球形，长7～8mm。花期6～9月，果期翌年2～5月。

Evergreen trees, to 25 m tall. Young branchlets terete, chaffy, grayish white or gray. Petiole 5-7 mm, leaf blade elliptic to ovate-elliptic, leathery. Inflorescences terminal or axillary on apical parts of branchlets, paniculate cymes, flower buds obovoid, 4-5 mm, petals distinct. Fruit subglobose, 7-8 mm. Fl. Jun.-Sep., fr. Feb.-May. of next year.

野外识别特征：

1. 小树皮灰白，光滑；老树皮褐色，粗糙；
2. 嫩枝圆形，有糠秕，灰白或灰色；
3. 叶椭圆形至卵状椭圆形，尾渐尖；
4. 花蕾倒卵形，果近球形，黑色。

Key notes for identification:

1. Bark gray, soomthly, old tree bark brown, rough;
2. Twigs terete, chaffy, grayish white or gray;
3. Leaf blade elliptic to ovate-elliptic, apex acute;
4. Flower buds obovoid, fruit subglobose, black.

花枝 Flowering branch
摄影：吴林芳 Photo by: Wu Linfang

果枝 Fruiting branch
摄影：吴林芳 Photo by: Wu Linfang

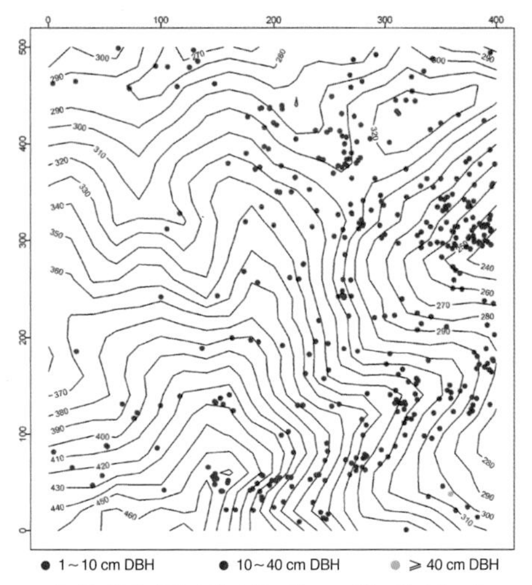

个体分布图 Distribution of individuals

径级分布表 DBH class

胸径等级 (Diameter class) (cm)	个体数 (No. of individuals in the plot)	比例 (Proportion) (%)
1～2	56	13.76
2～5	134	32.92
5～10	127	31.20
10～20	73	17.94
20～30	13	3.19
30～60	4	0.98
≥60	0	0.00

50 红枝蒲桃（红车） hóngzhīpútáo I Redbranche Syzygium

Syzygium rehderianum Merr. et Perry
桃金娘科 I Myrtaceae

代码（SpCode）= SYZREH
个体数（Individual number/20 hm^2）= 5917
最大胸径（Max DBH）= 51.0 cm
重要值排序（Importance value rank）= 3

常绿灌木或小乔木，高13m。嫩枝红色，圆形，稍压扁，老枝灰褐色。叶革质，椭圆形至狭椭圆形，叶柄较长，有7~9mm，边脉离边缘较远，约1~1.5mm。聚伞花序腋生，或生于枝顶叶腋内，花蕾长3.5mm，花瓣连合成帽状。果椭圆状卵形，长1.5~2cm，宽1cm。花期6~8月，果期11月至翌年1月。

Evergreen shrubs or small trees, to 13 m tall. Young branchlets red, terete, slightly compressed, old branches grayish brown. Petiole 7-9 mm, leaf blade elliptic to narrowly elliptic, leathery, intramarginal veins 1-1.5 mm from margin. Inflorescences axillary in axils apically on branches, cymes, flower buds ca. 3.5 mm, petals coherent. Fruit ellipsoid-ovoid, ca.1.5-2 cm × 1 cm. Fl. Jun.-Aug., fr. Nov.-Jan. of next year.

野外识别特征：

1. 树皮红褐色，局部粗糙；
2. 嫩枝红色，圆形，稍压扁；
3. 叶椭圆形至狭椭圆形，尾长渐尖，尖头5~10mm；
4. 花蕾倒圆锥形，果椭圆状卵形，黑色。

Key notes for identification:

1. Bark red-brown, part rough;
2. Twigs red, terete, slightly compressed;
3. Leaf blade elliptic to narrowly elliptic, apex long acuminate and with a 5-10 mm obtuse acumen;
4. Flower buds obconic, fruit ellipsoid-ovoid, black.

叶 Leaf
摄影：吴林芳 Photo by: Wu Linfang

叶背及树干 Leaf abaxial surface & trunk
摄影：吴林芳 Photo by: Wu Linfang

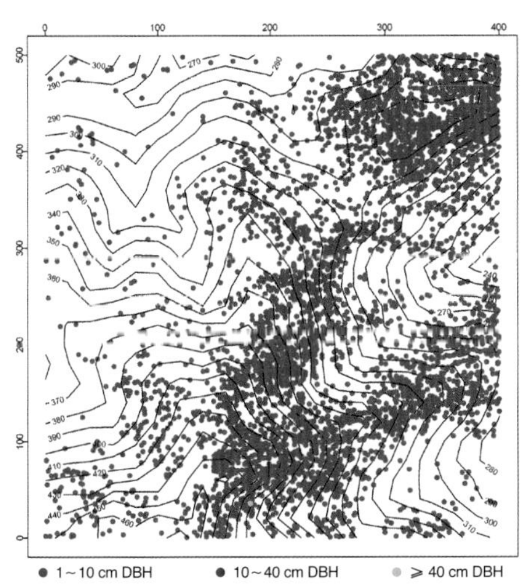

个体分布图 Distribution of individuals

径级分布表 DBH class

胸径等级 (Diameter class) (cm)	个体数 (No. of individuals in the plot)	比例 (Proportion) (%)
1~2	1230	20.79
2~5	2710	45.80
5~10	1405	23.75
10~20	539	9.11
20~30	31	0.52
30~60	2	0.03
≥60	0	0.00

51 柏拉木　bǎilāmù | Cochinchina Blastus

Blastus cochinchinensis Lour.
野牡丹科 | Melastomataceae

代码（SpCode）= BLACOC
个体数（Individual number/20 hm^2）= 3995
最大胸径（Max DBH）= 12.6 cm
重要值排序（Importance value rank）= 11

常绿灌木，高0.6～3m。茎圆柱形，幼时密被黄色腺点，以后脱落。叶纸质，披针形、狭椭圆形至椭圆状披针形，叶面被小腺点，后脱落，叶背密被小腺点，3(～5)基出脉。伞状聚伞花序，腋生，花白色至粉红色。蒴果椭圆形，4裂。花期6～8月，果期10～12月。

Evergreen shrubs, 0.6-3 m tall. Stems terete, densely yellow glandular when young, glabrescent. Leaf blade lanceolate, narrowly elliptic, to elliptic-lanceolate, densely glandular, adaxially sparsely glandular but glabrescent, 3(-5) basiveined. Inflorescences axillary, umbellate cymose, petals white to pink. Capsule elliptic, 4-sided. Fl. Jun.-Aug., fr. Oct.-Dec..

花及果　Flower & fruit
摄影：吴林芳　Photo by: Wu Linfang

叶背　Leaf abaxial surface
摄影：吴林芳　Photo by: Wu Linfang

叶　Leaf
摄影：吴林芳　Photo by: Wu Linfang

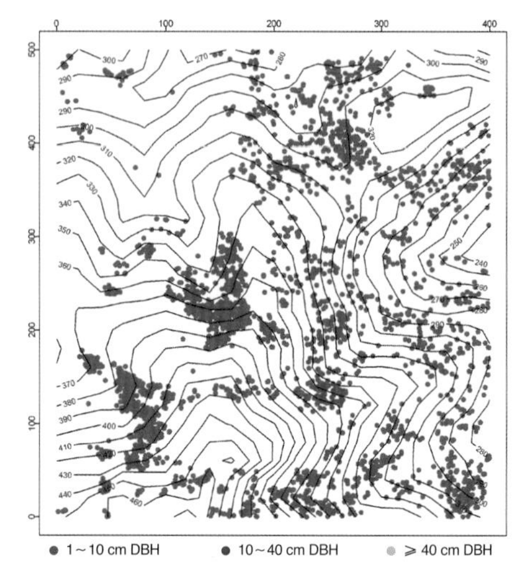

个体分布图 Distribution of individuals

径级分布表 DBH class

胸径等级 (Diameter class) (cm)	个体数 (No. of individuals in the plot)	比例 (Proportion) (%)
1～2	3018	75.54
2～5	971	24.31
5～10	6	0.15
10～20	0	0.00
20～30	0	0.00
30～60	0	0.00
≥60	0	0.00

52 毛菍 máoniè | Blood-red Melastoma

Melastoma sanguineum Sims
野牡丹科 | Melastomataceae

代码（SpCode）= MELSAN
个体数（Individual number/20 hm^2）= 97
最大胸径（Max DBH）= 7.9 cm
重要值排序（Importance value rank）= 68

常绿灌木，高1.5～3m。茎、小枝、叶柄、花梗及花萼均被平展的长粗毛，毛基部膨大。叶坚纸质，卵状披针形至披针形，全缘，基出脉5，两面被糙毛。伞房花序顶生，常仅1朵，有时3（～5）朵，花较大，花瓣粉红色或紫红色，长3～5cm。果杯状球形。花几全年，果期8～10月。

Evergreen shrubs 1.5-3 m tall. Stems, branches, petioles, pedicels and calyces densely hirsute, trichomes basally flattened. Leaf blade ovate-lanceolate to lanceolate, stiffly papery, margin entire, 5 basiveined. Inflorescences terminal, corymbose, usually 1 (or 3-5)-flowered, petals pink or purplish red, ca. 3-5 cm. Fruit urceolate-turbinate. Fl. almost all year, fr. Aug.-Oct..

果枝 Fruting branch
摄影：吴林芳 Photo by: Wu Linfang

花枝 Flowering branch
摄影：曹洪麟 Photo by: Cao Honglin

叶 Leaf
摄影：吴林芳 Photo by: Wu Linfang

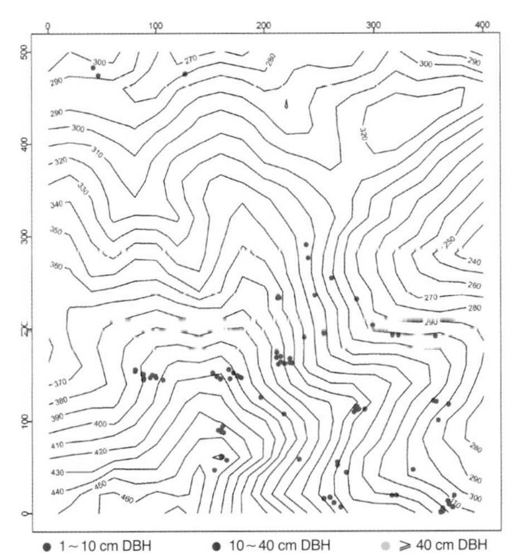

个体分布图 Distribution of individuals

径级分布表 DBH class

胸径等级 (Diameter class) (cm)	个体数 (No. of individuals in the plot)	比例 (Proportion) (%)
1～2	32	32.99
2～5	55	56.70
5～10	10	10.31
10～20	0	0.00
20～30	0	0.00
30～60	0	0.00
≥60	0	0.00

53 谷木 gǔmù | Privet Leaf Memecylon

Memecylon ligustrifolium Champ. ex Benth.
野牡丹科 | Melastomataceae

代码（SpCode）= MEMLIG
个体数（Individual number/20 hm^2）= 1288
最大胸径（Max DBH）= 36.3 cm
重要值排序（Importance value rank）= 20

常绿灌木或小乔木，高1.5～5（～7）m。小枝圆柱形或不明显的四棱形。叶革质，较大，长5.8～8cm，宽2.5～3.5cm，顶端渐尖，钝头，叶柄稍长，3～5mm。聚伞花序腋生，或生于落叶的叶腋，花瓣半圆形。果球形。花期5～8月，果期12月至翌年2月。

Evergreen shrubs or small trees, 1.5-5 (-7) m tall. Branches terete or sometimes 4-sided. Petiole 3-5 mm, leaf blade 5.8-8 cm × 2.5-3.5 cm, leathery, apex acuminate with an obtuse tip. Inflorescences in axils of leaves or at leaf scars on older branches, cymose, petals semiorbicular. Fruit a baccate drupe, globular. Fl. May-Aug., fr. Dec.-Feb. of next year.

野外识别特征：

1. 树皮褐色，具细纵裂；
2. 叶革质，顶端渐尖，具钝头；
3. 叶柄3～5mm。

Key notes for identification:

1. Bark brown, rough, longitudinally thinly fissured;
2. Leaf blade leathery, apex acuminate with an obtuse tip;
3. Petiole 3-5 mm.

花枝 Flowering branch
摄影：吴林芳 Photo by: Wu Linfang

果枝 Fruiting branch
摄影：吴林芳 Photo by: Wu Linfang

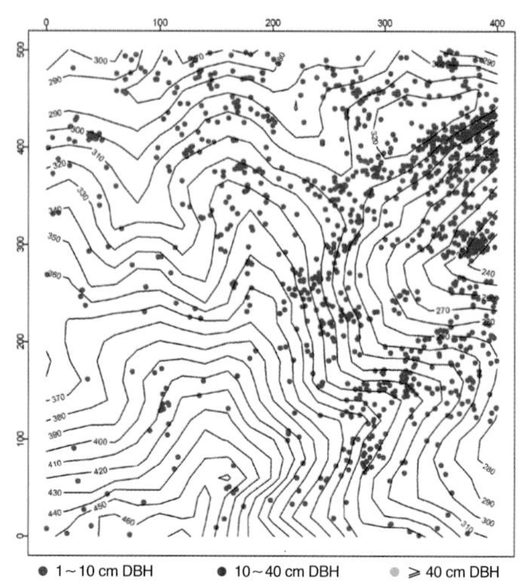

个体分布图 Distribution of individuals

径级分布表 DBH class

胸径等级 (Diameter class) (cm)	个体数 (No. of individuals in the plot)	比例 (Proportion) (%)
1～2	353	27.41
2～5	714	55.43
5～10	208	16.15
10～20	11	0.85
20～30	1	0.08
30～60	1	0.08
≥60	0	0.00

54 黑叶谷木 hēiyègǔmù | Black Leaf Memecylon

Memecylon nigrescens Hook. et Arn.
野牡丹科 | Melastomataceae

代码（SpCode）= MEMNIG
个体数（Individual number/20 hm^2）= 34
最大胸径（Max DBH）= 15 cm
重要值排序（Importance value rank）= 84

常绿灌木或小乔木，高2～8m。小枝圆柱形。叶坚纸质，较小，长3～6.5cm，宽1.5～3cm，顶端钝急尖，具微小尖头或有时微凹，叶柄较短，2～3mm。聚伞花序近头状，花瓣披针形。果球形，径6～7mm。花期5～6月，果期12月至翌年2月。

Evergreen shrubs or small trees, 2-8 m tall. Branches terete. Petiole 2-3 mm, leaf blade elliptic, rarely ovate-oblong, 3-6.5 cm × 1.5-3 cm, stiffly papery, apex obtuse to acute and sometimes with a retuse tip. Inflorescences axillary, cymose, nearly capitate, petals broadly lanceolate. Fruit baccate drupe, globular, 6-7 mm in diam.. Fl. May-Jun., fr. Dec.-Feb. of next year.

野外识别特征：
1. 树皮褐色，具细纵裂；
2. 叶坚纸质，顶钝急尖，具微小尖头或有时微凹；
3. 叶柄2～3mm。

Key notes for identification:
1. Bark brown, rough, longitudinally thinly fissured;
2. Leaf blade stiffly papery, apex obtuse to acute, with a retuse tip;
3. Petiole 2-3 mm.

果枝 Fruiting branch
摄影：吴林芳 Photo by: Wu Linfang

叶背 Leaf abaxial surface
摄影：吴林芳 Photo by: Wu Linfang

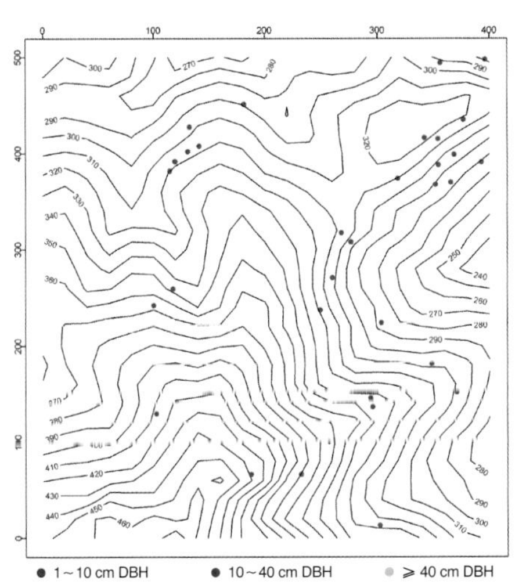

个体分布图 Distribution of individuals

径级分布表 DBH class

胸径等级 (Diameter class) (cm)	个体数 (No. of individuals in the plot)	比例 (Proportion) (%)
1～2	11	32.35
2～5	14	41.18
5～10	6	17.65
10～20	3	8.82
20～30	0	0.00
30～60	0	0.00
≥60	0	0.00

55 竹节树

zhújiéshù | India Carallia

Carallia brachiata (Lour.) Merr.
红树科 | Rhizophoraceae

代码（SpCode）= CARBRA
个体数（Individual number/20 hm²）= 639
最大胸径（Max DBH）= 24.1 cm
重要值排序（Importance value rank）= 30

常绿乔木，高7～10m。叶交互对生，全缘，稀具锯齿。聚伞花序腋生，花瓣白色，近圆形，边缘撕裂状。果实近球形，径4～5mm。花期冬季至翌年春季，果期春夏季。

Evergreen trees to 7-10 m tall. Leaves alternately opposite, leaf blade rarely, margin entire, serrate, or denticulate. Inflorescences axillary, cymose, petals white, suborbiculate, apically emarginate and unevenly lacerate. Fruit globose, ca. 4-5 mm in diam.. Fl. winter-spring of next year, fr. spring-summer.

野外识别特征：

1. 叶交互对生，全缘，稀具锯齿；
2. 托叶披针形，长1～2.5cm，生于叶间，早落，因此叶间具托叶痕。

Key notes for identification:

1. Leaves alternately opposite, margin entire or rarely serrate;
2. Stipules lanceolate, length 1-2.5 cm, interpetiolar, caducous, there are stipular scars interpetiolar.

果枝 Fruiting branch
摄影：吴林芳 Photo by: Wu Linfang

叶背 Leaf abaxial surface
摄影：吴林芳 Photo by: Wu Linfang

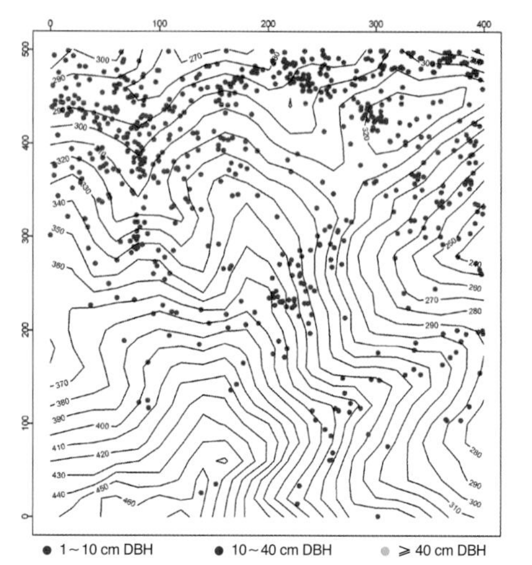

个体分布图 Distribution of individuals

径级分布表 DBH class

胸径等级 (Diameter class) (cm)	个体数 (No. of individuals in the plot)	比例 (Proportion) (%)
1～2	417	65.26
2～5	150	23.47
5～10	45	7.04
10～20	24	3.76
20～30	3	0.47
30～60	0	0.00
≥60	0	0.00

56 黄牛木

huángniúmù I Yellow Cow Wood

Cratoxylum cochinchinense (Lour.) Bl.
金丝桃科 I Hypericaceae

代码（SpCode）= CRACOC
个体数（Individual number/20 hm^2）= 7
最大胸径（Max DBH）= 13.5 cm
重要值排序（Importance value rank）= 118

落叶灌木或乔木，高1.5～18（～25）m。全体无毛。树干下部有簇生的长枝刺。树皮灰黄色或灰褐色。枝对生，幼枝略扁，淡红色。叶对生，坚纸质，椭圆形至长椭圆形或披针形。聚伞花序腋生、腋外生或顶生，有花（1～）2～3朵。蒴果椭圆形，棕色。花期4～5月，果期6月后。

Shrubs or trees, deciduous, 1.5-18 (-25) m tall. Glabrous. Trunk with clusters of long thorns on lower part. Bark gray-yellow or gray-brown. Twigs somewhat compressed, pink when young. Leaves opposite, leaf blade elliptic to oblong or lanceolate, stiffly papery. Cymes axillary or extra-axillary and terminal, (1 or) 2 or 3-flowered. Capsule brown, ellipsoid. Fl. Apr.-May, fr. after Jun..

树干 Trunk
摄影：吴林芳 Photo by: Wu Linfang

果枝 Fruiting branch
摄影：吴林芳 Photo by: Wu Linfang

花枝 Flowering branch
摄影：吴林芳 Photo by: Wu Linfang

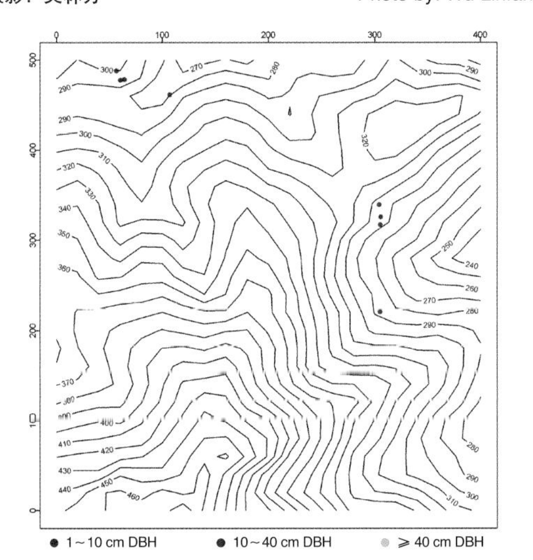

个体分布图 Distribution of individuals

径级分布表 DBH class

胸径等级 (Diameter class) (cm)	个体数 (No. of individuals in the plot)	比例 (Proportion) (%)
1～2	0	0.00
2～5	0	0.00
5～10	4	57.14
10～20	3	42.86
20～30	0	0.00
30～60	0	0.00
≥60	0	0.00

57 薄叶红厚壳（横经席） báoyèhónghòuké | Thin-leaved Calophyllum

Calophyllum membranaceum Gardn. et Champ.
藤黄科 | Clusiaceae

代码（SpCode）= CALMEM
个体数（Individual number/20 hm²）= 26
最大胸径（Max DBH）= 4.2 cm
重要值排序（Importance value rank）= 92

常绿灌木至小乔木，高1～5m。幼枝4棱具狭翅。叶对生，薄革质，侧脉纤细而密集，成规则的横行排列，边缘反卷。聚伞花序腋生，有花1～5(通常3)朵，花梗无毛。果卵状长圆球形，熟时黄色。花期3～5月，果期8～10(12)月。

Evergreen shrubs to small trees, 1-5 m tall. Young shoots tetragonous, narrowly winged. Leaves opposite, leaf blade thinly leathery, margin revolute, secondary veins many, almost perpendicular to midvein. Cyme axillary and terminating short axillary shoots, (1-) 3 (-5) -flowered, pedicels glabrous. Mature fruit yellow, ovoid-oblong. Fl. Mar.-May, fr. Aug.-Oct.(-Dec.).

野外识别特征：

1. 叶侧脉纤细而密集，成规则的横行排列；
2. 叶交互对生；
3. 幼枝4棱具狭翅。

Key notes for identification:

1. Leaf blade secondary veins many, almost perpendicular to midvein;
2. Leaves alternately opposite;
3. Twigs tetragonous, narrowly winged.

叶 Leaf
摄影：吴林芳 Photo by: Wu Linfang

果 Fruit
摄影：吴林芳 Photo by: Wu Linfang

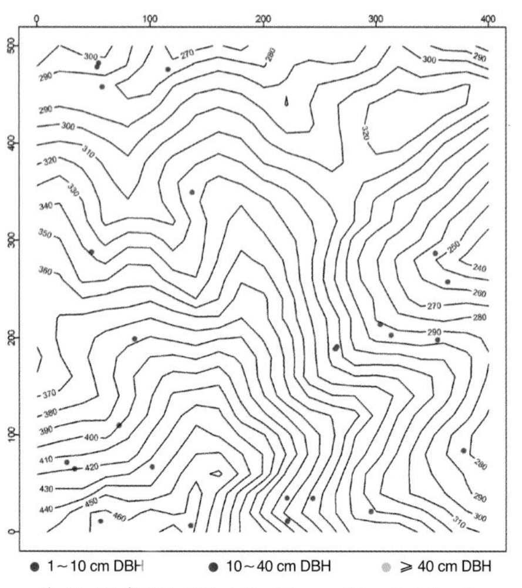

个体分布图 Distribution of individuals

径级分布表 DBH class

胸径等级 (Diameter class) (cm)	个体数 (No. of individuals in the plot)	比例 (Proportion) (%)
1～2	17	65.38
2～5	9	34.62
5～10	0	0.00
10～20	0	0.00
20～30	0	0.00
30～60	0	0.00
≥60	0	0.00

58 木竹子（多花山竹子） mùzhúzǐ | Many-flowered Garcinia

Garcinia multiflora Champ.ex Benth.
藤黄科 | Clusiaceae

代码（SpCode）= GARMUL
个体数（Individual number/20 hm^2）= 10
最大胸径（Max DBH）= 15.5 cm
重要值排序（Importance value rank）= 112

常绿乔木，高(3～)5～15m。树皮灰白色，粗糙。小枝绿色，具纵槽纹。叶对生，革质，全缘，卵形，长圆状卵形或长圆状倒卵形，边缘微反卷。花杂性，同株。果卵圆形至倒卵圆形，熟时黄色。花期6～8月，果期11～12月。

Evergreen trees, (3-) 5-15 m tall. Bark grey, scabrid. Twigs green, angled. Leaves opposite, leaf blade ovate, oblong-ovate, or oblong-obovate, leathery, margin entire, margin somewhat recurved. Plant monoecious. Mature fruit yellow, ovoid to obovoid. Fl. Jun.-Aug., fr. Nov.-Dec..

野外识别特征：

1. 叶交互对生，革质，全缘；
2. 无托叶，叶间无托叶痕；
3. 树皮灰白色；
4. 叶略大，长7～16（～20）cm，宽3～6cm。

Key notes for identification:

1. Leaves alternately opposite, margin entire, leathery;
2. No stipules, there are no stipular scars interpetiolar;
3. Bark gray;
4. Leaf blade 7-16 (-20) cm × 3-6 cm.

花 Flower
摄影：吴林芳 Photo by: Wu Linfang

果枝 Fruiting branch
摄影：吴林芳 Photo by: Wu Linfang

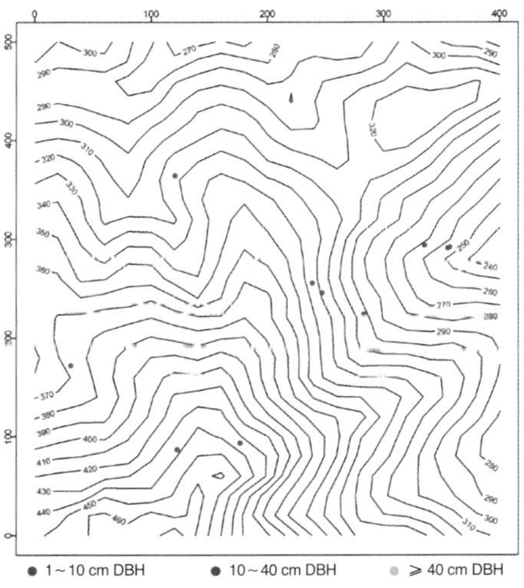

个体分布图 Distribution of individuals

径级分布表 DBH class

胸径等级 (Diameter class) (cm)	个体数 (No. of individuals in the plot)	比例 (Proportion) (%)
1～2	7	70.00
2～5	2	20.00
5～10	0	0.00
10～20	1	10.00
20～30	0	0.00
30～60	0	0.00
≥60	0	0.00

59 岭南山竹子

lǐngnánshānzhúzǐ | Lingnan Garcinia

Garcinia oblongifolia Champ. ex Benth.
藤黄科 | Clusiaceae

代码（SpCode）= GAROBL
个体数（Individual number/20 hm^2）= 612
最大胸径（Max DBH）= 37.0 cm
重要值排序（Importance value rank）= 28

常绿乔木或灌木，高5～15m。树皮深灰色，老枝通常具断环纹。叶对生，近革质，全缘，长圆形、倒卵状长圆形至倒披针形。花小，单性，异株，单生或成伞形状聚伞花序。浆果卵球形或圆球形。花期4～5月，果期10～12月。

Evergreen trees or shrubs, 5-15 m tall. Bark dark gray, branchlets usually with interrupted rings. Leaves opposite, leaf blade oblong, obovate-oblong to oblanceolate, margin entire, subleathery. Plant dioecious, flowers solitary or in an umbel-like cyme. Fruit ovoid or globose. Fl. Apr.-May, fr. Oct.-Dec.

野外识别特征：

1. 叶交互对生，近革质，全缘；
2. 无托叶，叶间无托叶痕；
3. 树皮深灰色；
4. 叶较小，长5～10cm，宽2～3.5cm。

Key notes for identification:

1. Leaves alternately opposite, margin entire, subleathery;
2. No stipules, there are no stipular scars interpetiolar;
3. Bark dark gray;
4. Leaf blade 5-10 cm × 2-3.5 cm.

叶 Leaf
摄影：吴林芳 Photo by: Wu Linfang

果枝 Fruiting branch
摄影：董安强 Photo by: Dong Anqiang

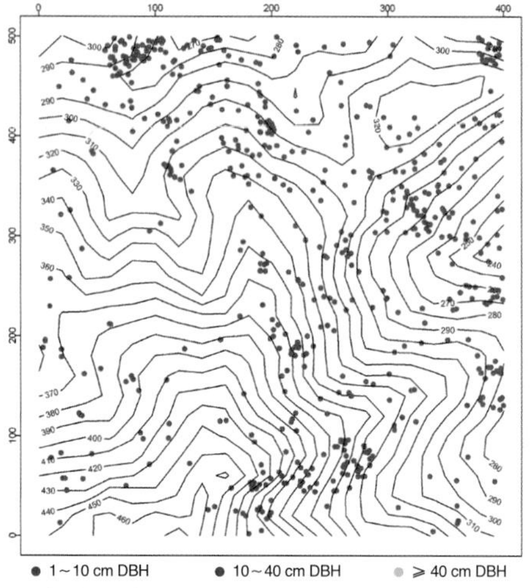

个体分布图 Distribution of individuals

径级分布表 DBH class

胸径等级 (Diameter class) (cm)	个体数 (No. of individuals in the plot)	比例 (Proportion) (%)
1～2	185	30.23
2～5	205	33.50
5～10	147	24.02
10～20	71	11.60
20～30	3	0.49
30～60	1	0.16
≥60	0	0.00

60 破布叶（布渣叶） pòbùyè I Microcos

Microcos paniculata Linn.
椴树科 I Tiliaceae

代码（SpCode）= MICPAN
个体数（Individual number/20 hm^2）= 3
最大胸径（Max DBH）= 15.3 cm
重要值排序（Importance value rank）= 145

落叶灌木或小乔木，高3～13m。树皮粗糙，嫩枝有毛。叶薄革质，卵状长圆形，两面初时有毛后秃净，托叶线状披针形。圆锥花序顶生，子房无毛。核果近球形或倒卵形，长约1cm。花期6～7月。

Deciduous shrubs or small trees 3-13 m tall. Bark rough, young branchlets hairy. Stipule filiform lanceolate, leaf blade ovate-oblong, thinly leathery, very sparsely stellate at first and glabrescent both abaxially and adaxially. Panicles terminal, ovary globose. Drupe nearly globose or obovoid, ca. 1 cm. Fl. Jun.-Jul..

果枝 Fruiting branch
摄影：吴林芳 Photo by: Wu Linfang

花 Flower
摄影：吴林芳 Photo by: Wu Linfang

叶 Leaf
摄影：吴林芳 Photo by: Wu Linfang

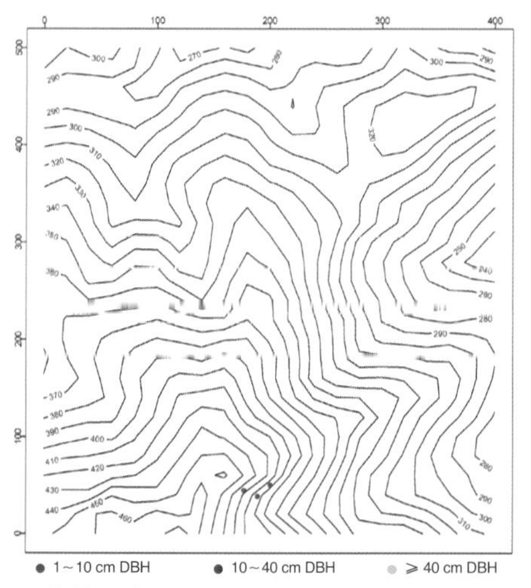

个体分布图 Distribution of individuals

径级分布表 DBH class

胸径等级 (Diameter class) (cm)	个体数 (No. of individuals in the plot)	比例 (Proportion) (%)
1～2	0	0.00
2～5	0	0.00
5～10	1	33.33
10～20	2	66.66
20～30	0	0.00
30～60	0	0.00
≥60	0	0.00

61 华杜英 huádùyīng | China Elaeocarpus

Elaeocarpus chinensis Hook ex Benth.
杜英科 | Elaeocarpaceae

代码（SpCode）= ELACHI
个体数（Individual number/20 hm^2）= 1
最大胸径（Max DBH）= 5.5 cm
重要值排序（Importance value rank）= 176

常绿小乔木，高7m。嫩枝有柔毛，老枝秃净。叶纸质，卵状披针形或披针形，5～8cm×2～3cm，初时有毛，很快秃净，叶背具细小黑腺点；叶柄长1.5～2cm，纤细。花瓣4，长圆形，近全缘，花药无芒刺。核果椭圆形，长不到1cm。花期5～6月，果期6～9月。

Evergreen small trees, to 7 m tall. Young branchlets puberulent, glabrous when old. Petiole 1.5-2 cm, slender, leaf blade ovate-lanceolate or lanceolate, 5-8 cm × 2-3 cm, papery, pubescent when young, soon glabrescent, abaxially black glandular punctate. Petals 4, oblong, margin nearly entire, anthers without hairs at apices. Drupe ellipsoid, shorter than 1 cm. Fl. May-Jun., fr. Jun.-Sep..

野外识别特征：

1. 嫩枝有柔毛，老枝秃净；
2. 叶纸质，卵状披针形或披针形；
3. 叶较小，长5～8cm，宽2～3cm；
4. 叶初时有毛，后秃净。

Key notes for identification:

1. Twigs puberulent, glabrous when old;
2. Leaf blade ovate-lanceolate or lanceolate papery;
3. Leaf blade relatively small, 5-8 × 2-3 cm;
4. Leaves pubescent when young, soon glabrescent.

果枝 Fruiting branch
摄影：吴林芳 Photo by: Wu Linfang

叶 Leaf
摄影：吴林芳 Photo by: Wu Linfang

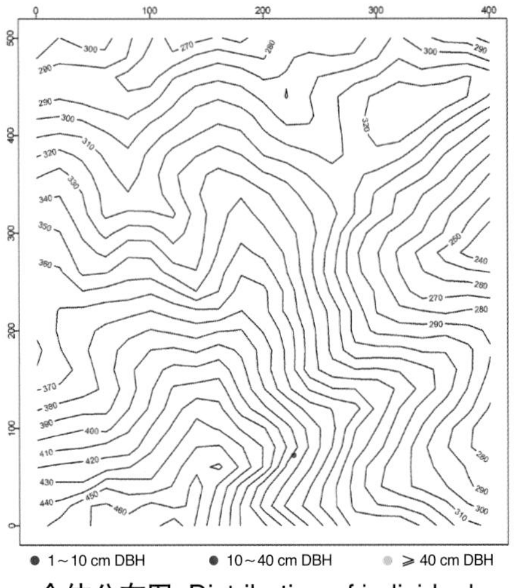

个体分布图 Distribution of individuals

径级分布表 DBH class

胸径等级 (Diameter class) (cm)	个体数 (No. of individuals in the plot)	比例 (Proportion) (%)
1～2	0	0.00
2～5	0	0.00
5～10	1	100.00
10～20	0	0.00
20～30	0	0.00
30～60	0	0.00
≥60	0	0.00

62 杜英

dùyīng I Common Elaeocarpus

Elaeocarpus decipiens Hemsl.
杜英科 I Elaeocarpaceae

代码（SpCode）= ELADEC
个体数（Individual number/20 hm^2）= 2
最大胸径（Max DBH）= 17.7 cm
重要值排序（Importance value rank）= 161

常绿乔木，高5～15m。嫩枝及顶芽初时被微毛，后秃净。叶革质，披针形或倒披针形，叶背秃净；叶柄长1cm，不膨大。总状花序多腋生或生于无叶的去年枝上，花药顶无芒刺。核果椭圆形，较大，长2～3.5cm，宽1.5～2cm，内果皮表面多数沟纹。花期6～7月。果期11月至翌年1月。

Evergreen trees, 5-15 m tall. Young branchlets and terminal buds puberulent, glabrescent. Petiole 1 cm, not swollen at upper end, puberulent, glabrescent, leaf blade blanceolate or oblanceolate, leathery, glabrous. Racemes in axils of fallen leaves, anthers no awn. Drupe ellipsoid, 2-3.5 cm × 1.5-2 cm, endocarp bony, prominently verrucose. Fl. Jun.-Jul., fr. Nov.-Jan. of next year.

野外识别特征：

1. 嫩枝被微毛，后秃净；
2. 叶革质，披针形或倒披针形；
3. 叶略长，长7～13.5cm，宽2～4cm；
4. 叶两面无毛。

Key notes for identification:

1. Twigs puberulent, glabrous when old;
2. Leaf blade oblanceolate or lanceolate leathery or thickly papery;
3. Leaf blade slightly longer, 7-13.5 cm × 2-4 cm;
4. Leaf blade glabrous.

果枝 Fruiting branch
摄影：董安强 Photo by: Dong Anqiang

叶背 Leaf abaxial surface
摄影：吴林芳 Photo by: Wu Linfang

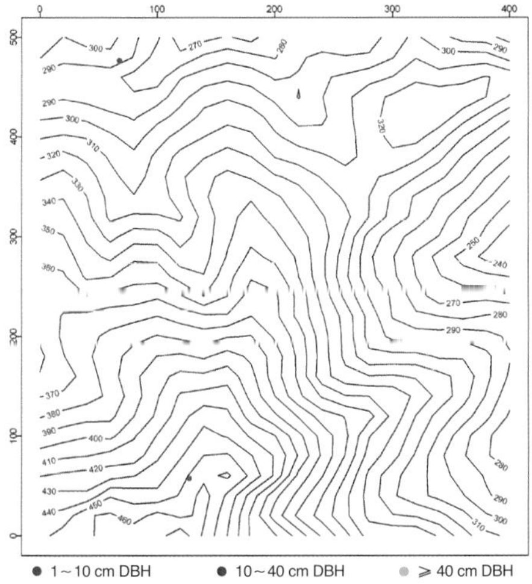

个体分布图 Distribution of individuals

径级分布表 DBH class

胸径等级 (Diameter class) (cm)	个体数 (No. of individuals in the plot)	比例 (Proportion) (%)
1～2	0	0.00
2～5	0	0.00
5～10	1	50.00
10～20	1	50.00
20～30	0	0.00
30～60	0	0.00
≥60	0	0.00

63 显脉杜英（拟杜英） xiǎnmàidùyīng | Mock Elaeocarpus

Elaeocarpus dubius A. DC.
杜英科 | Elaeocarpaceae

代码（SpCode）= ELADUB
个体数（Individual number/20 hm^2）= 344
最大胸径（Max DBH）= 30.4 cm
重要值排序（Importance value rank）= 37

常绿乔木，高达25m。嫩枝纤细，初时有银灰色短柔毛，后秃净。叶聚生枝顶，薄革质，长圆形或披针形，无毛，叶柄两端膨大。总状花序生于枝顶的叶腋内，花药顶具芒刺。核果椭圆形，小，长1～1.3cm，内果皮有腹缝沟。花期3～4月，果期4～6月。

Trees evergreen, to 25 m tall. Young branchlets slender, silvery-gray pubescent at first, glabrescent. Leaves crowded at twig apices, petiole, swollen at both ends, leaf blade oblong or lanceolate, thinly leathery, glabrous. Racemes in axils of fallen and current leaves, anthers with awn. Drupe ellipsoid, 1-1.3 cm long, endocarp bony. Fl. Mar.-Apr., fr. Apr.-Jun..

野外识别特征：

1. 嫩枝纤细，初时有银灰色短柔毛，后秃净；
2. 叶薄革质，长圆形或披针形，无毛；
3. 叶柄两端膨大。

Key notes for identification:

1. Branchlets slender, silvery-gray pubescent at first, glabrescent;
2. Leaf blade oblong or lanceolate, thinly leathery, glabrous;
3. Petiole swollen at both ends.

叶 Leaf
摄影：吴林芳 Photo by: Wu Linfang

叶背 Leaf abaxial surface
摄影：吴林芳 Photo by: Wu Linfang

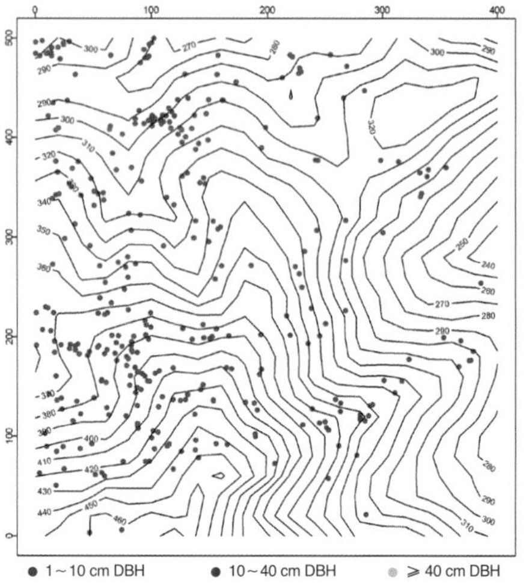

个体分布图 Distribution of individuals

径级分布表 DBH class

胸径等级 (Diameter class) (cm)	个体数 (No. of individuals in the plot)	比例 (Proportion) (%)
1～2	146	42.44
2～5	157	45.64
5～10	28	8.14
10～20	8	2.33
20～30	4	1.16
30～60	1	0.29
≥60	0	0.00

64 薯豆（日本杜英） shǔdòu | Japanese Elaeocarpus

Elaeocarpus japonicus Siet. & Zucc.
杜英科 | Elaeocarpaceae

代码（SpCode）= ELAJAP
个体数（Individual number/20 hm^2）= 4
最大胸径（Max DBH）= 17.1 cm
重要值排序（Importance value rank）= 138

常绿乔木，高25 m。嫩枝秃净无毛，叶芽有发亮绢毛。叶革质，通常卵形，有时椭圆形或倒卵形，初时有毛很快秃净，叶背有多数细小黑腺点；叶柄2～6 cm，两端常膨大。总状花序生于当年枝的叶腋内，花药顶无芒刺。核果小，椭圆形，长1～1.3 cm，内果皮无沟纹。花期4～5月，果期5～7月。

Evergreen trees to 25 m tall. Current year branchlets glabrous, buds sericeous. Petiole 2-6 cm, usually swollen at both ends, leaf blade ovate, elliptic, or obovate, leathery, both surfaces densely silvery-gray sericeous at first, soon glabrescent, abaxially with minute black glandular spots. Racemes axillary on branches of current year, anthers not awned at apices. Drupe shiny, ellipsoid, with no ventral sutures, 1-1.3 cm long. Fl. Apr.-May, fr. May-Jul..

野外识别特征：

1. 嫩枝秃净无毛；
2. 叶革质，卵形、椭圆形、倒卵形或披针形；
3. 叶初时有毛很快秃净；
4. 叶柄长，2～6cm，两端常膨大。

Key notes for identification:

1. Branchlets glabrous;
2. Leaf blade ovate, elliptic, obovate, lanceolate, leathery;
3. Leaf blade densely silvery-gray sericeous at first, soon glabrescent;
4. Petiole 2-6 cm, usually swollen at both ends.

果枝 Fruiting branch
摄影：吴林芳 Photo by: Wu Linfang

花枝 Flowering branch
摄影：吴林芳 Photo by: Wu Linfang

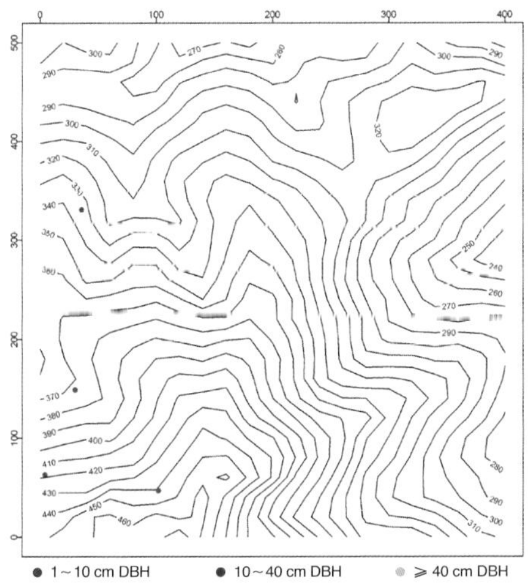

个体分布图 Distribution of individuals

径级分布表 DBH class

胸径等级 (Diameter class) (cm)	个体数 (No. of individuals in the plot)	比例 (Proportion) (%)
1～2	2	50.00
2～5	1	25.00
5～10	0	0.00
10～20	1	25.00
20～30	0	0.00
30～60	0	0.00
≥60	0	0.00

65 山杜英

shāndùyīng | Sylvestral Elaeocarpus

Elaeocarpus sylvestris Poir.
杜英科 | Elaeocarpaceae

代码（SpCode）= ELASYL
个体数（Individual number/20 hm^2）= 176
最大胸径（Max DBH）= 27.2 cm
重要值排序（Importance value rank）= 52

常绿小乔木，高15m。小枝纤细，常被微毛。叶纸质，不发亮，倒卵形或倒披针形，两面无毛；叶柄短，1～1.5cm，初时被毛，后秃净。花药顶无芒刺。核果小，椭圆形，长1～1.2cm，内果皮有3腹缝沟。花期4～5月，果期5～8月。

Evergreen small trees to 15 m tall. Branchlets slender, sparsely pale pilose. Petiole 1-1.5 cm, sparsely pubescent at first, glabrescent, leaf blade obovate or oblanceolate, papery, both surfaces glabrous. Anthers not awned at apices. Drupe ellipsoid, 1-1.2 cm long, endocarp thinly bony, with 3 ventral sutures. Fl. Apr.-May, fr. May-Aug..

野外识别特征：

1. 嫩枝纤细，常被微毛；
2. 叶纸质，倒卵形或倒披针形，无毛；
3. 叶两面无毛；
4. 叶柄初时被毛，后秃净。

Key notes for identification:

1. Branchlets slender, sparsely pale pilose;
2. Leaf blade obovate or oblanceolate, papery;
3. Leaf blade both surfaces glabrous;
4. Petiole sparsely pubescent at first, glabrescent.

花枝 Flowering branch
摄影：吴林芳 Photo by: Wu Linfang

叶背 Leaf abaxial surface
摄影：吴林芳 Photo by: Wu Linfang

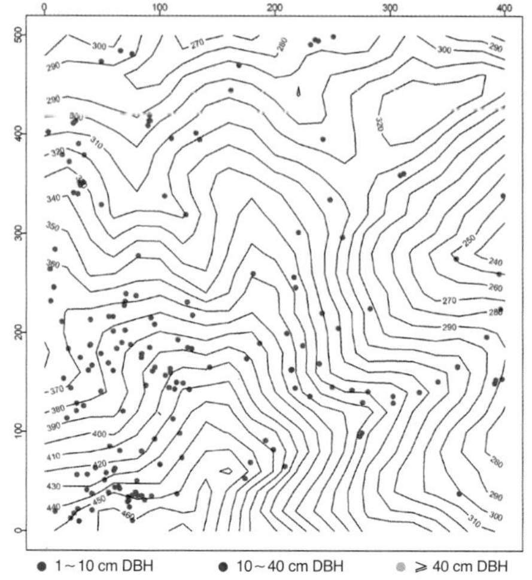

个体分布图 Distribution of individuals

径级分布表 DBH class

胸径等级 (Diameter class) (cm)	个体数 (No. of individuals in the plot)	比例 (Proportion) (%)
1～2	62	35.23
2～5	78	44.32
5～10	23	13.07
10～20	11	6.25
20～30	2	1.14
30～60	0	0.00
≥60	0	0.00

66 猴欢喜

hóuhuānxǐ | Chinese Sloanea

Sloanea sinensis (Hance) Hemsl.
杜英科 | Elaeocarpaceae

代码（SpCode）= SLOSIN
个体数（Individual number/20 hm^2）= 2
最大胸径（Max DBH）= 3.5 cm
重要值排序（Importance value rank）= 173

常绿乔木，高20m。嫩枝无毛。叶革质，形状多变，常长圆形至狭倒卵形，无毛，通常全缘，有时上半部有数个疏齿；叶柄1～4cm，无毛。花多朵簇生于枝顶叶腋。蒴果针刺较长，1～1.5cm。花期6～10月，果期翌年夏秋季。

Evergreen trees to 20 m tall. Young branchlets glabrous. Petiole 1-4 cm, glabrous, leaf blade variable, usually oblong or narrowly obovate, leathery, glabrous, margin usually entire, sometimes sparsely dentate along upper margin. Flowers fascicled in axils at twig apices. Capsule prickles 1-1.5 cm. Fl. Jun.-Oct., fr. summer-autumn of next year.

蒴果 Capsule
摄影：吴林芳 Photo by: Wu Linfang

叶 Leaf
摄影：吴林芳 Photo by: Wu Linfang

叶背 Leaf abaxial surface
摄影：吴林芳 Photo by: Wu Linfang

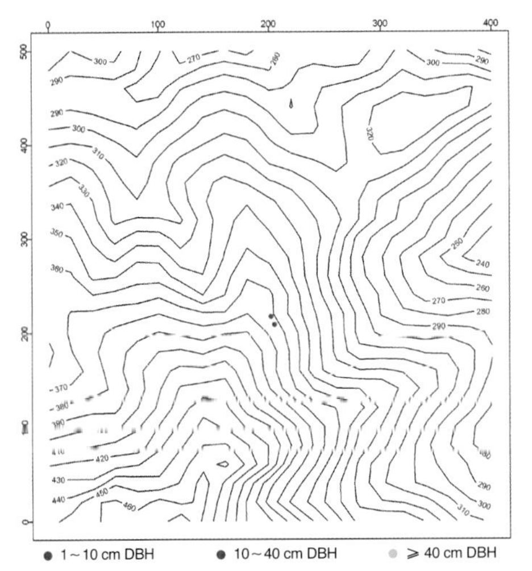

个体分布图 Distribution of individuals

径级分布表 DBH class

胸径等级 (Diameter class) (cm)	个体数 (No. of individuals in the plot)	比例 (Proportion) (%)
1～2	0	0.00
2～5	2	100.0
5～10	0	0.00
10～20	0	0.00
20～30	0	0.00
30～60	0	0.00
≥60	0	0.00

67 翻白叶树（半枫荷） fānbáiyèshù | Heterophyllous Wingseedtree

Pterospermum heterophyllum Hance
梧桐科 | Sterculiaceae

代码（SpCode）= PTEHET
个体数（Individual number/20 hm^2）= 48
最大胸径（Max DBH）= 43 cm
重要值排序（Importance value rank）= 89

半常绿乔木，高20m。叶二型，幼树或萌蘖枝上的叶盾形，掌状3～5裂；成年树叶矩圆形至卵状矩圆形，长7～15cm，宽3～10cm。花单生或2～4朵组成腋生聚伞花序。蒴果木质，矩圆状卵形，6×2～2.5cm，果柄粗壮而短，1～1.5cm。花期秋季。

Semievergreen trees, to 20 m tall. Leaves dimorphic, juvenile and coppice leaves: leaf blade palmately 3-5-lobed, base conspicuously peltate, mature tree leaves: leaf blade oblong-ovate to oblong, 7-15 cm × 3-10 cm. Flowers solitary or in cymes of 2-4, axillary. Capsule woody, cylindrical-ovoid, ca. 6 cm × 2-2.5 cm, base tapering into robust, 1-1.5 cm stipe. Fl. autumn.

蒴果 Capsule
摄影：吴林芳 Photo by: Wu Linfang

花枝 Flowering branch
摄影：吴林芳 Photo by: Wu Linfang

幼叶 Young leaf
摄影：吴林芳 Photo by: Wu Linfang

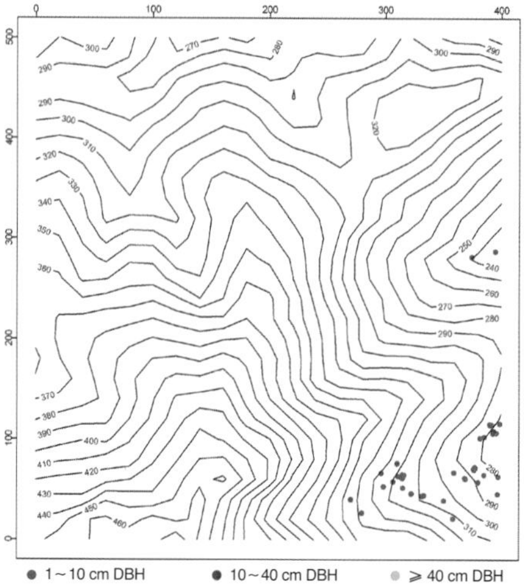

个体分布图 Distribution of individuals

径级分布表 DBH class

胸径等级 (Diameter class) (cm)	个体数 (No. of individuals in the plot)	比例 (Proportion) (%)
1～2	21	43.75
2～5	16	33.33
5～10	7	14.58
10～20	1	2.08
20～30	2	4.17
30～60	1	2.08
≥60	0	0.00

68 窄叶半枫荷

zhǎiyèbànfēnghé | Lanceleaf Wingseedtree

Pterospermum lanceifolium Roxb.
梧桐科 | Sterculiaceae

代码（SpCode）= PTELAN
个体数（Individual number/20 hm^2）= 3
最大胸径（Max DBH）= 5.5 cm
重要值排序（Importance value rank）= 155

半常绿乔木，高25m。叶披针形或矩圆状披针形，长5～9cm，宽2～3cm；叶柄长约5mm。花白色，单生于叶腋。蒴果木质，矩圆状卵形，长5cm，宽2cm；果柄纤弱，长3～5cm。花期春夏季。

Semi evergreen trees, to 25 m tall. Leaf blade lanceolate or oblong-lanceolate, 5-9 cm × 2-3 cm, petiole 5 mm. Flowers solitary, axillary, petals white. Capsule woody, cylindrical-ovoid, ca. 5 cm × 2 cm, base tapering into slender, 3-5 cm stipe. Fl. spring and summer.

野外识别特征：

1. 叶不二型；
2. 叶披针形或矩圆状披针形；
3. 叶较窄，长5～9cm，宽2～3cm；
4. 叶柄较短，仅5mm左右。

Key notes for identification:

1. Leaf blade not double-form;
2. Leaf blade lanceolate or oblong-lanceolate;
3. Leaf blade narrowly, 5-9 cm × 2-3 cm;
4. Petiole ca. 0.5 cm.

叶 Leaf
摄影：吴林芳 Photo by: Wu Linfang

叶背 Leaf abaxial surface
摄影：吴林芳 Photo by: Wu Linfang

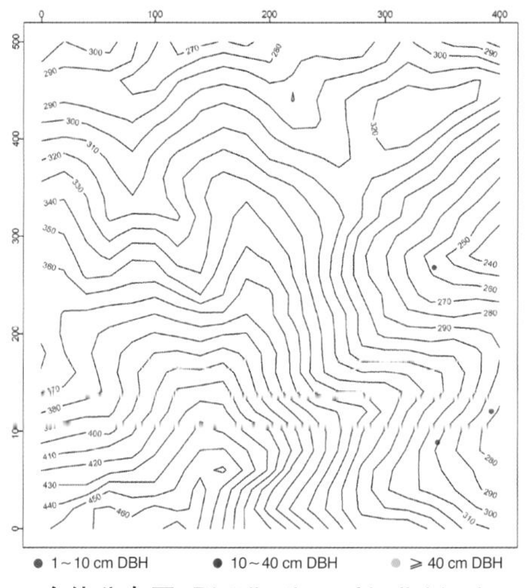

个体分布图 Distribution of individuals

径级分布表 DBH class

胸径等级 (Diameter class) (cm)	个体数 (No. of individuals in the plot)	比例 (Proportion) (%)
1～2	0	0.00
2～5	2	66.67
5～10	1	33.33
10～20	0	0.00
20～30	0	0.00
30～60	0	0.00
≥60	0	0.00

69 两广梭椤 liǎngguǎngsuōluó | Bunch-like Reevesia

Reevesia thyrsoidca Lindl.
梧桐科 | Sterculiaceae

代码（SpCode）= REETHY
个体数（Individual number/20 hm^2）= 181
最大胸径（Max DBH）= 24.0 cm
重要值排序（Importance value rank）= 61

常绿乔木。树皮灰褐色。幼枝干时棕黑色，略被稀疏的星状短柔毛。叶革质，矩圆形、椭圆形或矩圆状椭圆形，两面无毛；叶柄长1～3cm，两端膨大。聚伞状伞房花序顶生，花瓣5，匙形，白色。蒴果矩圆状梨形，有5棱。花期3～4月。

Evergreen, trees. Bark gray-brown. Young branchlets brownish black when dried, sparsely stellate puberulent. Petiole 1-3 cm, swollen at both ends, leaf blade oblong to elliptic, leathery, both surfaces glabrous. Inflorescence cymose-corymbose, densely flowered, petals 5, white, spatulate. Capsule oblong-pyriform, 5-angular. Fl. Mar.-Apr..

野外识别特征：

1. 树皮灰褐色，具皮孔；
2. 叶革质，两面无毛，先端急尖或渐尖；
3. 叶柄长1～3cm，两端膨大；
4. 叶中脉富纤维。

Key notes for identification:

1. Bark gray-brown, lenticellate;
2. Leaf blade leathery, both surfaces glabrous, apex acute or acuminate;
3. Petiole 1-3 cm, swollen at both ends;
4. Midvein rich in fibers.

枝叶 Branch and leaves
摄影：曹洪麟 Photo by: Cao Honglin

花序 Inflorescence
摄影：曹洪麟 Photo by: Cao Honglin

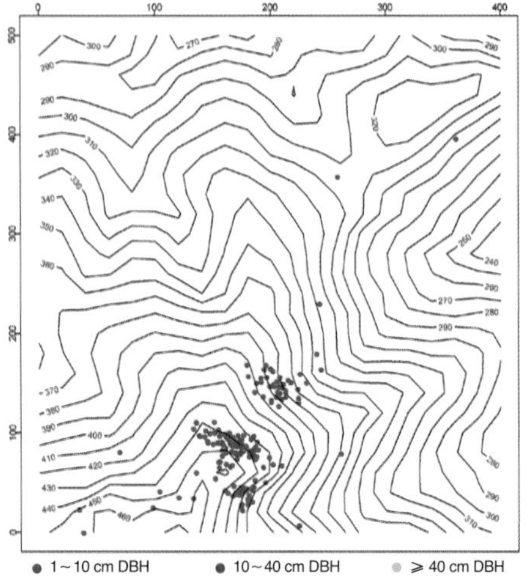

个体分布图 Distribution of individuals

径级分布表 DBH class

胸径等级 (Diameter class) (cm)	个体数 (No. of individuals in the plot)	比例 (Proportion) (%)
1～2	32	17.68
2～5	91	50.28
5～10	37	20.44
10～20	20	11.05
20～30	1	0.55
30～60	0	0.00
≥60	0	0.00

70 假苹婆

jiǎpíngpó | Lanceleaf Sterculia

Sterculia lanceolata Cav.
梧桐科 | Sterculiaceae

代码（SpCode）= STELAN
个体数（Individual number/20 hm^2）= 52
最大胸径（Max DBH）= 38 cm
重要值排序（Importance value rank）= 69

常绿乔木。小枝幼时被毛。叶椭圆形、披针形或椭圆状披针形；叶柄长2.5～3.5cm，两端膨大。圆锥花序腋生，长4～10cm，密集而多分枝，花萼淡红色。蓇葖果鲜红色，长卵形或长椭圆形；种子黑褐色，椭圆状卵形，约1cm。花期4～6月。

Evergreen trees. Branchlets at first pilose. Petiole 2.5-3.5 cm, swollen at both ends, leaf blade elliptic, lanceolate, or elliptic-lanceolate. Inflorescence paniculate, 4-10 cm, densely many-branched, calyx reddish, follicle red, narrowly ovoid or ellipsoid, seeds black-brown, ellipsoid-ovoid, ca. 1 cm. Fl. Apr.-Jun..

花序 Inflorescence
摄影：吴林芳 Photo by: Wu Linfang

种子 Seeds
摄影：吴林芳 Photo by: Wu Linfang

果枝 Fruiting branch
摄影：吴林芳 Photo by: Wu Linfang

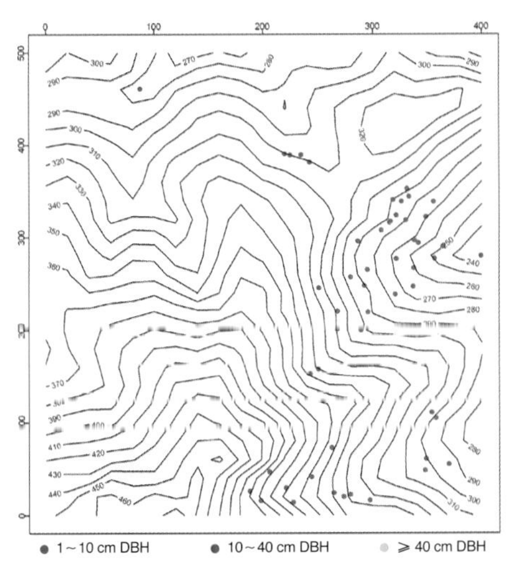

个体分布图 Distribution of individuals

径级分布表 DBH class

胸径等级 (Diameter class) (cm)	个体数 (No. of individuals in the plot)	比例 (Proportion) (%)
1～2	10	19.23
2～5	22	42.31
5～10	12	23.08
10～20	5	9.62
20～30	2	3.85
30～60	1	1.92
≥60	0	0.00

71 红背山麻杆（红背叶）

hóngbèishānmágān | Christmas Bush

Alchornea trewioides (Benth.) Muell.-Arg.
大戟科 | Euphorbiaceae

代码（SpCode）= ALCTRE
个体数（Individual number/20 hm^2）= 14
最大胸径（Max DBH）= 2.7 cm
重要值排序（Importance value rank）= 124

落叶灌木。叶薄纸质，阔卵形，叶面无毛，叶背浅红色，仅沿脉被柔毛，基部具斑状腺体4个，基出脉3条，小托叶披针形。雌雄异株，雄花序穗状，腋生或生于1年生已落叶腋部，长7～15cm。蒴果球形，径8～10mm。花期4～5月，果期6～8月。

Deciduous shrubs. Leaf blade broadly ovate, thinly papery, abaxially puberulent along veins, with 4 glands, stipels lanceolate, basal veins 3. Diecious, male inflorescences axillary, often at leafless nodes, unbranched, 7-15 cm. Capsule globose, 8-10 mm in diam.. Fl. Apr.-May, fr. Jun.-Aug..

果枝 Fruiting branch
摄影：吴林芳 Photo by: Wu Linfang

蒴果 Capsule
摄影：吴林芳 Photo by: Wu Linfang

花序 Inflorescence
摄影：吴林芳 Photo by: Wu Linfang

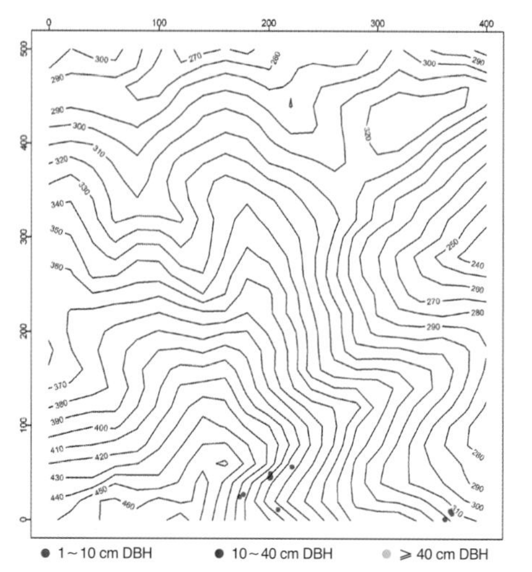

个体分布图 Distribution of individuals

径级分布表 DBH class

胸径等级 (Diameter class) (cm)	个体数 (No. of individuals in the plot)	比例 (Proportion) (%)
1～2	4	28.57
2～5	10	71.43
5～10	0	0.00
10～20	0	0.00
20～30	0	0.00
30～60	0	0.00
≥60	0	0.00

72 五月茶

wǔyuèchá | Chinese Laurel

Antidesma bunius (L.) Spreng
大戟科 | Euphorbiaceae

代码（SpCode）= ANTBUN
个体数（Individual number/20 hm^2）= 7
最大胸径（Max DBH）= 3.7 cm
重要值排序（Importance value rank）= 129

常绿乔木，罕为灌木，高达30m。小枝无毛或被疏柔毛。托叶线形，早落；叶革质或厚纸质，长圆形、椭圆形或倒卵形，无毛或有时叶背中脉有毛。花序腋生或顶生，雄花序为穗状；雌花为总状，子房无毛。核果近球形或椭圆形，径约8mm，熟时红色。花期3～5月，果期6～10月。

Evergreen trees, rarely shrubs, up to 30 m tall. Twigs glabrous to very shortly pubescent. Stipules linear, caduceus, leaf blade oblong, elliptic, or obovate, leathery or thickly papery, glabrous except sometimes midvein pilose. Inflorescences axillary or terminal, male inflorescences spicate, female inflorescences racemose, ovary glabrous. Drupes ellipsoid, ca. 8 mm, red to black when ripe. Fl. Mar.-May, fr. Jun.-Oct..

果枝 Fruiting branch
摄影：吴林芳 Photo by: Wu Linfang

叶背 Leaf abaxial surface
摄影：吴林芳 Photo by: Wu Linfang

花序 Inflorescence
摄影：吴林芳 Photo by: Wu Linfang

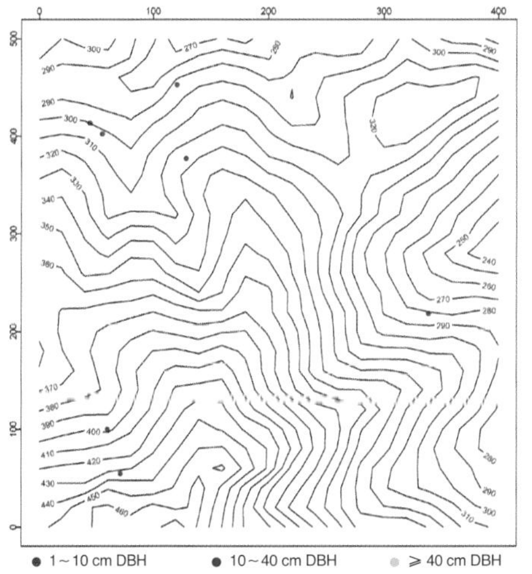

个体分布图 Distribution of individuals

径级分布表 DBH class

胸径等级 (Diameter class) (cm)	个体数 (No. of individuals in the plot)	比例 (Proportion) (%)
1～2	4	57.14
2～5	3	42.86
5～10	0	0.00
10～20	0	0.00
20～30	0	0.00
30～60	0	0.00
≥60	0	0.00

73 黄毛五月茶 huángmáowǔyuèchá | Salamander Tree

Antidesma fordii Hemsl.
大戟科 | Euphorbiaceae

代码（SpCode）= ANTFOR
个体数（Individual number/20 hm^2）= 1
最大胸径（Max DBH）= 5.0 cm
重要值排序（Importance value rank）= 179

常绿小乔木，高7m。枝条圆柱形，小枝、叶柄、托叶、花序轴被黄茸毛，其余被长柔毛或柔毛。托叶卵形至披针形；叶片长圆形、椭圆形或倒卵形，纸质，叶柄长1～3cm。花序顶生或腋生，子房被毛，雄花萼片(3)4～6。核果纺锤形。花期3～7月，果期7月至翌年1月。

Evergreen small trees up to 7 m tall. Twigs, petioles, and inflorescence axes densely yellow tomentose, other parts densely villous or pubescent. Stipules ovate to lanceolate, petiole 1-3 cm, leaf blade oblong, sometimes elliptic, slightly ovate or obovate, papery. Inflorescences terminal and axillary, ovary pubescent, male flower sepals (3) 4-6. Drupes ellipsoid, laterally compressed. Fl. Mar.-Jul., fr. Jul.-Jan.of next year.

果序 Infructescence
摄影：吴林芳 Photo by: Wu Linfang

叶 Leaf
摄影：吴林芳 Photo by: Wu Linfang

叶背 Leaf abaxial surface
摄影：吴林芳 Photo by: Wu Linfang

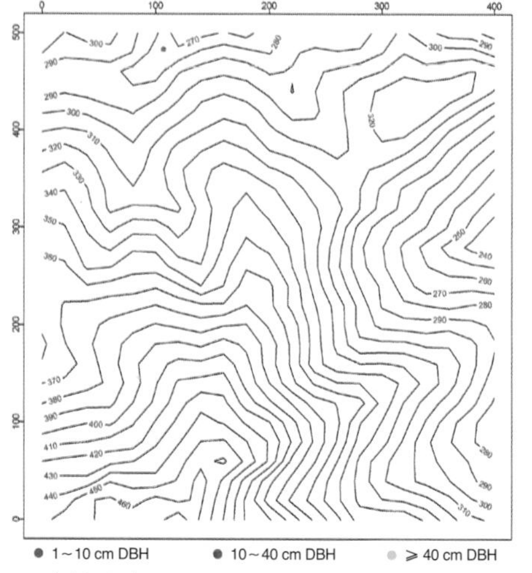

个体分布图 Distribution of individuals

径级分布表 DBH class

胸径等级 (Diameter class) (cm)	个体数 (No. of individuals in the plot)	比例 (Proportion) (%)
1～2	0	0.00
2～5	1	100.00
5～10	0	0.00
10～20	0	0.00
20～30	0	0.00
30～60	0	0.00
≥60	0	0.00

74 酸味子

suānwèizǐ | Japanese China Laurel

Antidesma japonicum Sieb. et Zucc.
大戟科 | Euphorbiaceae

代码（SpCode）= ANTJAP
个体数（Individual number/20 hm^2）= 105
最大胸径（Max DBH）= 18.0 cm
重要值排序（Importance value rank）= 60

半常绿乔木或灌木，高2～8m。小枝初时被短柔毛，后变无毛。叶片纸质至近革质，椭圆形、长椭圆形至长圆状披针形，叶面无光泽，托叶线形早落。总状花序顶生或腋生，子房无毛，花盘垫状。核果椭圆形，5～6mm，花期4～8月，果期6～9月。

Semievergreen shrubs or trees, 2-8 m tall. Twigs delicate, pubescent when young, glabrescent. Stipules linear, caduceus, leaf blade elliptic or oblong-elliptic to oblong-lanceolate, papery to subleathery, adaxially not shiny. Inflorescences terminal or axillary, ovary glabrous, disk cushion-shaped. Drupes ellipsoid, 5-6 mm. Fl. Apr.-Aug., fr. Jun.-Sep..

野外识别特征：

1. 小枝初时被短柔毛，后秃净；
2. 叶纸质至近革质，椭圆形、长椭圆形至长圆状披针形，渐尖至尾尖；
3. 叶背仅中脉披短柔毛。

Key notes for identification:

1. Twigs pubescent when young, glabrescent;
2. Leaf blade papery to subleathery, elliptic oblong-elliptic to oblong-lanceolate, apex acuminate or acute, sometimes caudate;
3. Leaves glabrous except for midvein sometimes pubescent.

果枝 Fruiting branch
摄影：吴林芳 Photo by: Wu Linfang

叶 Leaf
摄影：吴林芳 Photo by: Wu Linfang

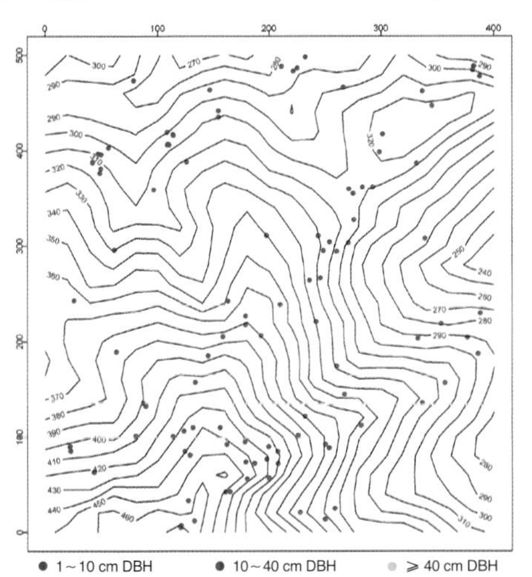

个体分布图 Distribution of individuals

径级分布表 DBH class

胸径等级 (Diameter class) (cm)	个体数 (No. of individuals in the plot)	比例 (Proportion) (%)
1～2	59	56.19
2～5	41	39.05
5～10	4	3.81
10～20	1	0.95
20～30	0	0.00
30～60	0	0.00
≥60	0	0.00

75 小叶五月茶

xiǎoyèwǔyuèchá | Small-leaf China Laurel

Antidesma montanum var. *microphyllum* (Hemsl.) Petra Hoffmann

大戟科 | Euphorbiaceae

代码（SpCode）= ANTMON

个体数（Individual number/20 hm^2）= 42

最大胸径（Max DBH）= 7.5 cm

重要值排序（Importance value rank）= 78

半常绿灌木，高2～4m。小枝圆柱形，密被黄色茸毛，后变无毛。叶近革质，狭披针形或狭长圆状椭圆形；托叶线状披针形。总状花序单个或2～3个聚生于枝顶或叶腋内；花盘杯状，子房无毛。核果卵圆形，紫黑色。花期5～6月，果期6～11月。

Semievergreen shrubs, 2-4 m tall. Branchlets densely yellow-pubescent, glabrescent. Leaves subleathery, stipules linear-lanceolate, laef blade narrowly lanceolate or narrowly oblong-elliotic. Racemes solitary or 2-3-clustered, pubescent, disk annular, ovary glabrous. Drupes ovoid, blackish purple. Fl. May-Jun., fr. Jun.-Nov..

野外识别特征：

1. 小枝密被黄色茸毛后变无毛；
2. 叶近革质，披针形至线形，先端钝或渐尖；
3. 叶背被微毛或仅叶脉被毛或无毛。

Key notes for identification:

1. Branchlets densely yellow pubescent, glabrescent;
2. Leaves subleathery, lanceolate, or linear apex obtuse or acuminate;
3. Leaves abaxially slightly pubescent throughout, or only vines pubescent,or glabrous.

叶 Leaf
摄影：吴林芳 Photo by: Wu Linfang

叶背及果枝 Leaf abaxial surface & fruiting branch
摄影：吴林芳 Photo by: Wu Linfang

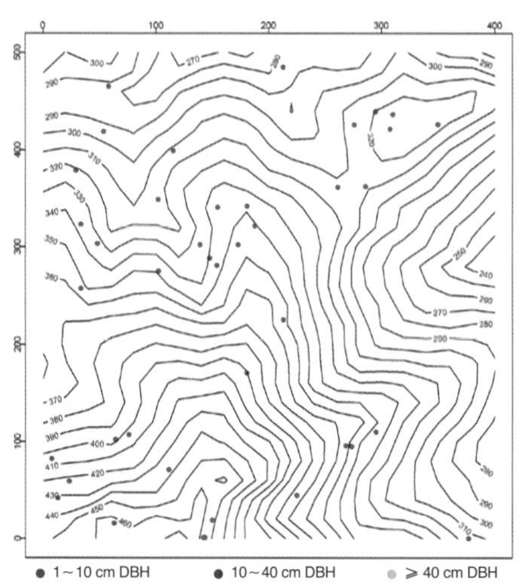

个体分布图 Distribution of individuals

径级分布表 DBH class

胸径等级 (Diameter class) (cm)	个体数 (No. of individuals in the plot)	比例 (Proportion) (%)
1～2	23	54.76
2～5	17	40.48
5～10	2	4.76
10～20	0	0.00
20～30	0	0.00
30～60	0	0.00
≥60	0	0.00

76 云南银柴

yúnnányínchái | Yunnan Aporosa

Aporosa yunnanensis (Pax & K. Hoffmann) Metc.
大戟科 | Euphorbiaceae

代码（SpCode）= APOYUN
个体数（Individual number/20 hm^2）= 3722
最大胸径（Max DBH）= 17 cm
重要值排序（Importance value rank）= 9

常绿小乔木，高8 m。幼枝光滑无毛。托叶早落，叶柄长1～1.3cm，顶端各1个腺体，叶片长圆形、长椭圆形或长卵形至披针形。花单性，雌雄异株，穗状花序，子房无毛。蒴果近圆形，熟时红色，无毛。花果期1～10月。

Evergreen small trees up to 8 m tall. Young branches smooth, glabrous. Stipules caducous, petiole 1-1.3 cm, apex bilateral with 2 glands, leaf blade oblong, oblong-elliptic, or oblong-ovate to lanceolate. Flower unisexual, dioecism, spica, ovary glabrous. Capsules subglobose, red when mature, glabrous. Fl. and fr. Jan.-Oct..

蒴果 Capsule
摄影：曹洪麟 Photo by: Cao Honglin

叶 Leaf
摄影：吴林芳 Photo by: Wu Linfang

花序 Inflorescence
摄影：曹洪麟 Photo by: Cao Honglin

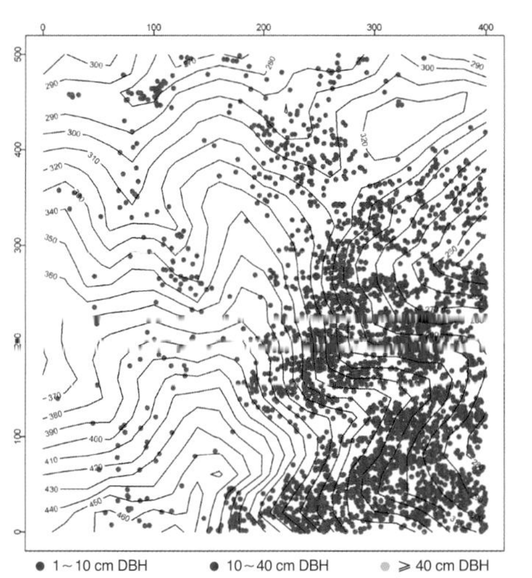

个体分布图 Distribution of individuals

径级分布表 DBH class

胸径等级 (Diameter class) (cm)	个体数 (No. of individuals in the plot)	比例 (Proportion) (%)
1～2	238	6.39
2～5	1899	51.02
5～10	1484	39.87
10～20	101	2.71
20～30	0	0.00
30～60	0	0.00
≥60	0	0.00

77 大叶土蜜树

dàyètǔmìshù | Ford's Bridelia

Bridelia retusa (Linn.) A.Juss.
大戟科 | Euphorbiaceae

代码（SpCode）= BRIFOR
个体数（Individual number/20 hm²）= 80
最大胸径（Max DBH）= 31.0 cm
重要值排序（Importance value rank）= 64

落叶乔木，高达15m。小树基部有枝刺，小枝、叶片、叶柄均无毛。叶片纸质，倒卵形，有时长圆形，叶柄长1.2cm，稍粗壮，托叶早落。花小，雌雄异株，穗状花序腋生或3～9穗花组成顶生圆锥花序。核果卵形，2室，黑色。花期4～9月，果期8月至翌年1月。

Deciduous trees up to 15 m tall. Branches, leaf blade and petiole are glabrous. Leaf blade obovate, sometimes elliptic, papery, petiole 1.2 cm, slightly stout, stipules caduceus. Inflorescences many flowered, grouped into axillary spikes or 3-9-spiked and grouped into panicles at apex of branchlets, dioecism. Drupes ovoid or depressed globose, black, 2-celled. Fl. Apr.-Sep., fr. Aug.-Jan.of next year.

树干 Trunk
摄影：吴林芳 Photo by: Wu Linfang

果枝 Fruiting branch
摄影：董安强 Photo by: Dong Anqiang

叶背 Leaf abaxial surface
摄影：吴林芳 Photo by: Wu Linfang

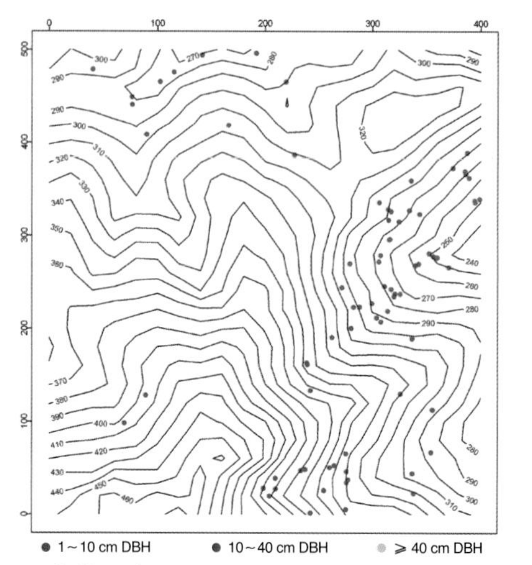

个体分布图 Distribution of individuals

径级分布表 DBH class

胸径等级 (Diameter class) (cm)	个体数 (No. of individuals in the plot)	比例 (Proportion) (%)
1～2	37	46.25
2～5	28	35.00
5～10	5	6.25
10～20	4	5.00
20～30	5	6.25
30～60	1	1.25
≥60	0	0.00

78 毛果巴豆 máoguǒbādòu I Hairyfruit Croton

Croton lachnocarpus Benth.
大戟科 I Euphorbiaceae

代码（SpCode）= CROLAC
个体数（Individual number/20 hm^2）= 57
最大胸径（Max DBH）= 4.1 cm
重要值排序（Importance value rank）= 116

常绿灌木，高1～3m。1年生枝、幼叶、花序和果密被星状柔毛，老枝近无毛。叶纸质，长圆形、长圆状椭圆形至椭圆状卵形，边缘具不明确细齿，基出脉3，叶基具2腺体。总状花序1～3个顶生，苞片钻形。蒴果稍扁球形，被毛。花期4～5月，果期6～9月。

Evergreen shrubs to 1-3 m tall. Indumentum densely stellate-pubescent, older branches subglabrous. Leaf blade oblong-elliptic to elliptic-ovate, papery, margin obscurely serrulate, petiole apex or base of leaf blade with 2 stalked and cupular glands, basal veins 3. Inflorescences 1-3, terminal, bracts subulate. Capsules slightly oblate, hairy. Fl. Apr.-May, fr. Jun.-Sep..

叶背 Leaf abaxial surface
摄影：吴林芳 Photo by: Wu Linfang

果枝 Fruiting branch
摄影：吴林芳 Photo by: Wu Linfang

花序 Inflorescence
摄影：吴林芳 Photo by: Wu Linfang

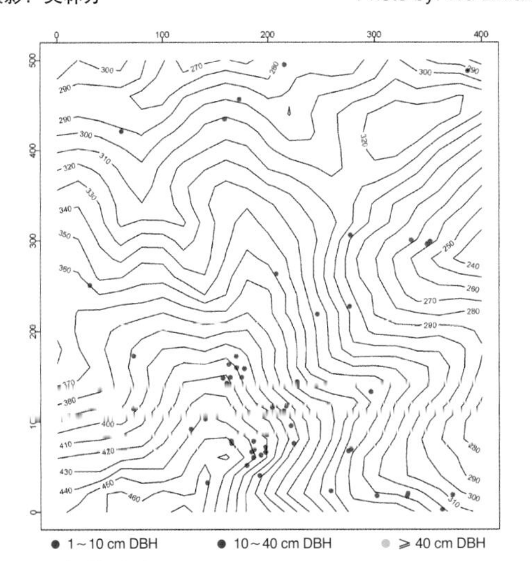

个体分布图 Distribution of individuals

径级分布表 DBH class

胸径等级 (Diameter class) (cm)	个体数 (No. of individuals in the plot)	比例 (Proportion) (%)
1～2	40	70.18
2～5	17	29.82
5～10	0	0.00
10～20	0	0.00
20～30	0	0.00
30～60	0	0.00
≥60	0	0.00

79 毛果算盘子 máoguǒsuànpánzǐ | Hairy-fruited Abacus Plant

Glochidion eriocarpum Champ.ex Benth.
大戟科 | Euphorbiaceae

代码（SpCode）= GLOERI
个体数（Individual number/20 hm^2）= 9
最大胸径（Max DBH）= 2.8 cm
重要值排序（Importance value rank）= 73

常绿灌木，高5m。雌雄同株；小枝密被长柔毛。叶二列，两面均被长柔毛，纸质，卵形，狭或宽，叶基两侧对称。花单生或2~4朵簇生叶腋内，雄蕊3，花柱3倍长于子房。蒴果扁球形，密被长柔毛。花果期几全年。

Evergreen shrubs to 5 m tall. Monoecious, branchlets densely spreading villous. Leaves distichous, both sides densely villous, papery leaf blade ovate, narrowly or broadly, base symmetrical. Flowers axillary, solitary or in 2-4-flowered clusters, stamens 3, style column 3 times length then ovary. Capsules depressed globose, densely villous. Fl. and fr. almost throughout year.

野外识别特征：

1. 小枝密被长柔毛；
2. 叶二列，两面密被长柔毛；
3. 叶基两侧对称。

Key notes for identification:

1. Branchlets densely spreading villous;
2. Leaves distichous, both sides densely spreading villous;
3. Leaf blade base symmetrical.

花枝 Flowering branch
摄影：吴林芳 Photo by: Wu Linfang

蒴果 Capsule
摄影：吴林芳 Photo by: Wu Linfang

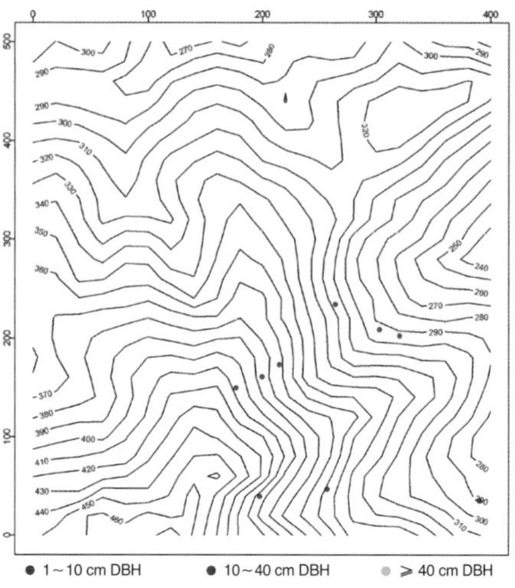

个体分布图 Distribution of individuals

径级分布表 DBH class

胸径等级 (Diameter class) (cm)	个体数 (No. of individuals in the plot)	比例 (Proportion) (%)
1~2	7	77.78
2~5	2	22.22
5~10	0	0.00
10~20	0	0.00
20~30	0	0.00
30~60	0	0.00
≥60	0	0.00

80 白背算盘子

báibèisuànpánzǐ | Wright's Abacus Plant

Glochidion wrightii Benth.
大戟科 | Euphorbiaceae

代码（SpCode）= GLOWRI
个体数（Individual number/20 hm^2）= 19
最大胸径（Max DBH）= 7.5 cm
重要值排序（Importance value rank）= 98

常绿灌木或乔木，高1～8m。雌雄同株，全株无毛。叶纸质，长圆形或长圆状披针形，常呈镰刀状弯斜，叶背干后灰白色。雌雄花同簇生于叶腋内，雄蕊3，花柱合生成圆柱状。蒴果扁球形，直径8～15mm，红色。花期4～8月，果期7～11月。

Evergreen shrubs or trees to 1-8 m tall. Monoecious, glabrous throughout. Leaf blade oblong or oblong-lanceolate, often obliquely falcate, papery, abaxially gray-glaucous when dry, base inequilateral, Flowers in bisexual clusters, stamens 3, style column annular. Capsules depressed-globose, 8-15 mm in diam., reddish when mature. Fl. Apr.-Aug., fr. Jul.-Nov..

野外识别特征：

1. 全株无毛；
2. 叶纸质，常呈镰刀状弯斜；
3. 叶面绿色，叶背粉绿色。

Key notes for identification:

1. Glabrous throughout;
2. Leaf blade papery often obliquely falcate;
3. Leaf blade green adaxially, farinose-greenish abaxially.

叶 Leaf
摄影：吴林芳 Photo by: Wu Linfang

叶背 Leaf abaxial surface
摄影：吴林芳 Photo by: Wu Linfang

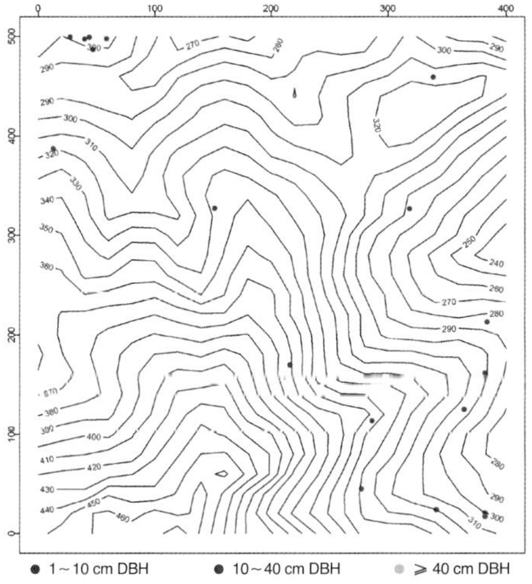

个体分布图 Distribution of individuals

径级分布表 DBH class

胸径等级 (Diameter class) (cm)	个体数 (No. of individuals in the plot)	比例 (Proportion) (%)
1～2	7	36.84
2～5	7	36.84
5～10	5	26.32
10～20	0	0.00
20～30	0	0.00
30～60	0	0.00
≥60	0	0.00

81 轮苞血桐（安德曼血桐） lúnbāoxuětóng | Andaman's Macaranga

Macaranga andamanica Kurz
大戟科 | Euphorbiaceae

代码（SpCode）= MACAND
个体数（Individual number/20 hm^2）= 25
最大胸径（Max DBH）= 14.1 cm
重要值排序（Importance value rank）= 107

常绿灌木，高1～5m。小枝被短毛，后变无毛。叶柄2～4cm，被毛；叶厚纸质，长圆状披针形或长圆形，相对较小，长7～14cm，宽2.5～5.5cm，浅盾状或非盾状着生，叶基具2腺体。雌花通常1朵，有时雌雄同株，同时具雄花1～2朵。蒴果双球形，长5～11mm，宽12mm。花果期几全年。

Evergreen shrubs 1-5 m tall. Branchlets pubescent, glabrescent. Petiole 2-4 cm, pilose, leaf blade oblong-lanceolate or oblong, 7-14 cm × 2.5-5.5 cm, thickly papery, base narrowly peltate or not, with 2 glands. Female inflorescences often 1-flowered, sometimes inflorescence bisexual, with 1 or 2 male flowers. Capsule 2-globed, ca. 5-11 mm × 12 mm. Fl. and fr. almost year-round.

野外识别特征：

1. 叶互生，区别于粗毛野桐；
2. 小枝被短毛，后变无毛；
3. 叶柄长2～4cm；
4. 花腋生。

Key notes for identification:

1. Leaves alternate, distinguish from *Hancea hookeriana*;
2. Branchlets pubescent, glabrescent;
3. Petiole 2-4 cm;
4. Inflorescences axillary.

叶 Leaf
摄影：吴林芳 Photo by: Wu Linfang

叶背 Leaf abaxial surface
摄影：吴林芳 Photo by: Wu Linfang

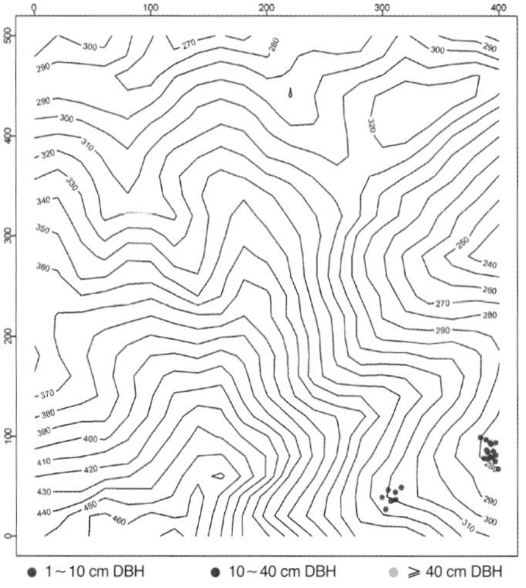

个体分布图 Distribution of individuals

径级分布表 DBH class

胸径等级 (Diameter class) (cm)	个体数 (No. of individuals in the plot)	比例 (Proportion) (%)
1～2	6	24.00
2～5	7	28.00
5～10	5	20.00
10～20	7	28.00
20～30	0	0.00
30～60	0	0.00
≥60	0	0.00

82 鼎湖血桐

dǐnghúxuětóng | Sampson Macarange

Macaranga sampsonii Hance
大戟科 | Euphorbiaceae

代码（SpCode）= MACSAM
个体数（Individual number/20 hm^2）= 673
最大胸径（Max DBH）= 40.4 cm
重要值排序（Importance value rank）= 41

常绿灌木或小乔木，高2～7m。林窗中的先锋树种。叶薄革质，三角状卵形或卵圆形，浅盾状着生，侧脉约7对，托叶披针形，被毛，早落。圆锥花序，苞片卵状披针形，雄蕊4（3～5）枚。蒴果双球形。花期5～6月，果期7～8月。

Evergreen small trees or shrubs, 2-7 m tall. A pioneerspecies of the gap succession. Stipules lanceolate, pubescent, deciduous, leaf blade deltoid-ovate or orbicular-ovate, thinly leathery, base narrowly peltate, lateral veins about 7 pairs. Male and female inflorescences paniculate, bracts ovate-lanceolate, stamens (3 or) 4 (or 5). Capsule 2-lobed. Fl. May-Jun., fr. Jul.-Aug..

果序 Infructescence
摄影：吴林芳 Photo by: Wu Linfang

叶 Leaf
摄影：吴林芳 Photo by: Wu Linfang

花序 Inflorescence
摄影：吴林芳 Photo by: Wu Linfang

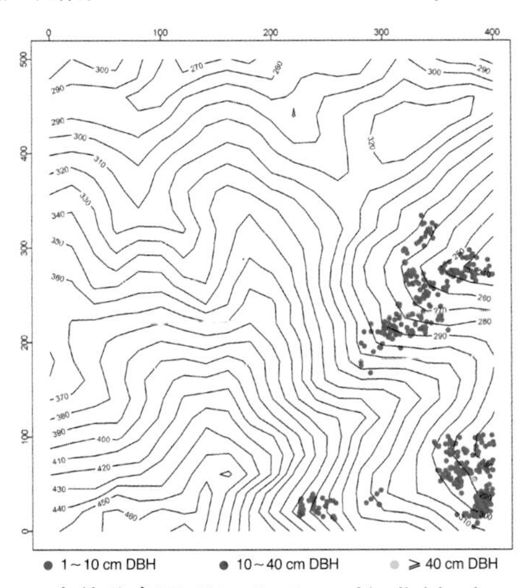

个体分布图 Distribution of individuals

径级分布表 DBH class

胸径等级 (Diameter class) (cm)	个体数 (No. of individuals in the plot)	比例 (Proportion) (%)
1～2	201	29.87
2～5	335	49.78
5～10	109	16.20
10～20	25	3.71
20～30	2	0.30
30～60	1	0.15
≥60	0	0.00

83 粗毛野桐 cūmáoyětóng | Hooker Mallotus

Hancea hookeriana Seem.
大戟科 | Euphorbiaceae

代码（SpCode）= HANHOO
个体数（Individual number/20 hm^2）= 15
最大胸径（Max DBH）= 5 cm
重要值排序（Importance value rank）= 132

常绿灌木或小乔木，高1.5～6m。小枝和叶柄被疏生黄色长粗毛。叶对生，但同对的叶型和大小极不相同；小型叶退化成托叶状，正常叶长披针形，边缘近全缘或波状。花雌雄异株，雄花序总状，雌花单生。蒴果三棱状球形，具刺被毛。花期3～5月，果期8～10月。

Evergreen shrubs or small trees, 1.5-6 m tall. Branchlets and petioles sparsely hispid with stiff spreading hairs. Leaves opposite, differing in size and shape, normally oblong-lanceolate but sometimes reduced and stipule-shaped, margin subentire or undulate. Flowers dioecious, male racemes and female solitary. Capsules triangular-globose, with soft prickles. Fl. Mar.-May, fr. Aug.-Oct..

野外识别特征：

1. 叶对生，但小叶型退化成托叶状；
2. 小叶被疏生黄色长粗毛；
3. 叶柄长1～1.5cm；
4. 花顶生。

Key notes for identification:

1. Leaves opposite, small leaf blade stipule-shaped;
2. Young branchlets sparsely hispid with stiff spreading hairs;
3. Petiole 1-1.5 cm;
4. Inflorescences terminal.

叶 Leaf
摄影：吴林芳 Photo by: Wu Linfang

叶背及蒴果 Leaf abaxial surface & capsule
摄影：吴林芳 Photo by: Wu Linfang

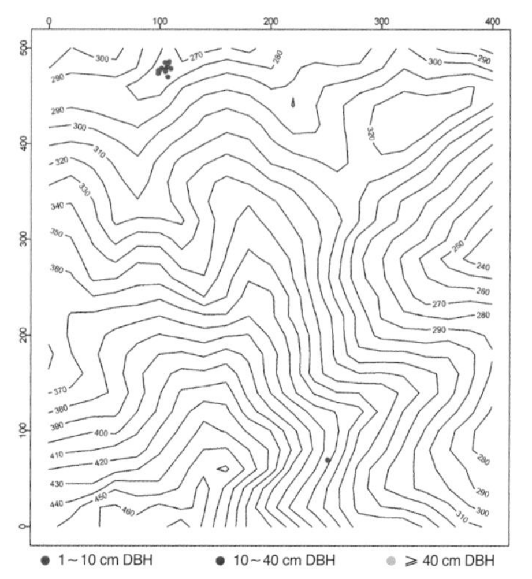

个体分布图 Distribution of individuals

径级分布表 DBH class

胸径等级 (Diameter class) (cm)	个体数 (No. of individuals in the plot)	比例 (Proportion) (%)
1～2	5	33.33
2～5	9	60.00
5～10	1	6.67
10～20	0	0.00
20～30	0	0.00
30～60	0	0.00
≥60	0	0.00

84 白背叶

báibèiyè I Whitebackleaf Mallotus

Mallotus apelta (Lour.) Muell.-Arg.
大戟科 I Euphorbiaceae

代码（SpCode）= MALAPE
个体数（Individual number/20 hm^2）= 4
最大胸径（Max DBH）= 12.3 cm
重要值排序（Importance value rank）= 85

常绿灌木或小乔木，高1～6m。叶厚纸质，互生，大致卵形，基出脉3条，叶背被灰白色星状茸毛。花雌雄异株，雄花序圆锥或穗状花序，雄蕊50～75；雌花序穗状，长15～60cm。蒴果近球形，密被线形软刺。花期5～9月，果期8～11月。

Evergreen shrubs or small trees 1-6 m tall. Leaf blade alternate, broadly ovate, thickly papery, basal veins 3, abaxially whitish tomentulose. Male inflorescences terminal, stamens 50-75, female infructescence 15-60 cm. Capsule subglobose, densely softly spiny, spines filiform. Fl. May-Sep., fr. Aug.-Nov..

野外识别特征：

1. 蒴果的软刺线形，长3～8mm；
2. 叶不盾状着生；
3. 叶卵形或阔卵形。

Key notes for identification:

1. Capsule's softly spiny filiform, 3-8 mm;
2. Leaves base not peltate;
3. Leaf blade ovate or broadly ovate.

花序 Inflorescence
摄影：吴林芳 Photo by: Wu Linfang

叶背及果序 Leaf abaxial surface & infructescence
摄影：吴林芳 Photo by: Wu Linfang

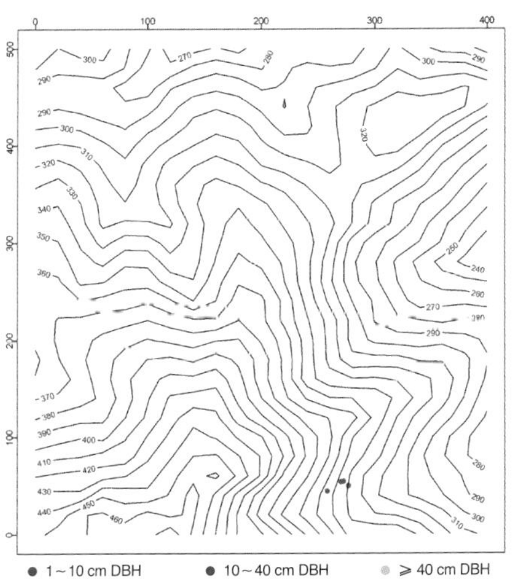

个体分布图 Distribution of individuals

径级分布表 DBH class

胸径等级 (Diameter class) (cm)	个体数 (No. of individuals in the plot)	比例 (Proportion) (%)
1～2	1	25.00
2～5	0	0.00
5～10	2	50.00
10～20	1	25.00
20～30	0	0.00
30～60	0	0.00
≥60	0	0.00

85 白楸

báiqīu | Panicled Mallotus

Mallotus paniculatus (Lam.) Müell. -Arg.

大戟科 | Euphorbiaceae

代码（SpCode）= MALPAN

个体数（Individual number/20 hm^2）= 155

最大胸径（Max DBH）= 23.0 cm

重要值排序（Importance value rank）= 59

常绿灌木或小乔木，高3～15m。叶互生，卵形、卵状三角形或菱形，基出脉3～5条，稍盾状着生，叶背被白色星状茸毛。花雌雄异株，圆锥花序或总状花序，雄蕊30～50，苞片卵形。蒴果扁球形，具3个分果室，被稀疏软刺，软刺锥形，长4～5mm。花期7～10月，果期10～12月。

Evergreen shrubs or small trees, 3-15 m tall. Leaf blade alternate, rhombic, ovate, or ovate-triangular, often 1-3-lobed or 3-cuspidate, basal veins 3-5, abaxially grayish tomentulose. Flowers dioecious, male and female inflorescences often branched, stamens 30-50; bracts ovate. Capsule depressed-globose, 3-locular, sparsely softly spiny, spines subulate, 4-5 mm. Fl. Jul.-Oct., fr. Oct.-Dec..

野外识别特征：

1. 蒴果的软刺钻形，长4～5mm；
2. 叶稍盾状着生；
3. 叶菱形、卵状或三角状卵形。

Key notes for identification:

1. Capsule's softly spiny subulate, 4-5 mm;
2. Leaf blade base narrowly peltate;
3. Leaf blade rhombic, ovate, or triangular-ovate.

花序及果枝 Inflorescence & fruiting branch

摄影：吴林芳 Photo by: Wu Linfang

叶背 Leaf abaxial surface

摄影：吴林芳 Photo by: Wu Linfang

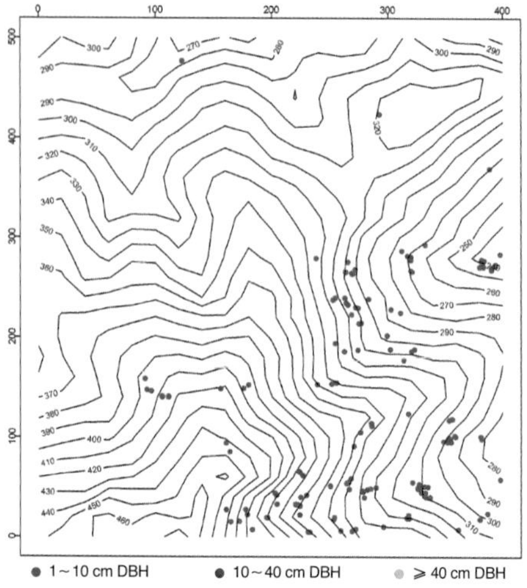

个体分布图 Distribution of individuals

径级分布表 DBH class

胸径等级 (Diameter class) (cm)	个体数 (No. of individuals in the plot)	比例 (Proportion) (%)
1～2	37	23.87
2～5	80	51.61
5～10	23	14.84
10～20	13	8.39
20～30	2	1.29
30～60	0	0.00
≥60	0	0.00

86 山乌桕 shānwūjiù I Mountain Tallow Tree

Triadica cochichinensis Lour.
大戟科 I Euphorbiaceae

代码（SpCode）= TRICOC
个体数（Individual number/20 hm^2）= 24
最大胸径（Max DBH）= 6.7 cm
重要值排序（Importance value rank）= 95

落叶小乔木，高5～12m。全株无毛，具白色乳汁。叶纸质，椭圆形或长圆状卵形，叶柄2～7.5cm，顶端具2个腺体。总状花序顶生，基部为数朵雌花。蒴果球形，熟时黑色，种子具白色蜡质层。花期4～6月，果期7～10月。

Deciduous small trees, 5-12 m tall. Glabrous throughout, with milky juice. Leaves papery, petioles 2-7.5 cm, with 2 glands at apex, leaf blade elliptic or oblong-ovate. Racemes terminal, with several female flowers at base, and many male flowers on upper part. Capsules black when mature, globose, seeds with waxy aril. Fl. Apr.-Jun. fr. Jul.-Oct..

幼苗 Seedling
摄影：吴林芳 photo by: Wu Linfang

花枝 Flowering branch
摄影：吴林芳 Photo by: Wu Linfang

果枝 Fruiting branch
摄影：吴林芳 Photo by: Wu Linfang

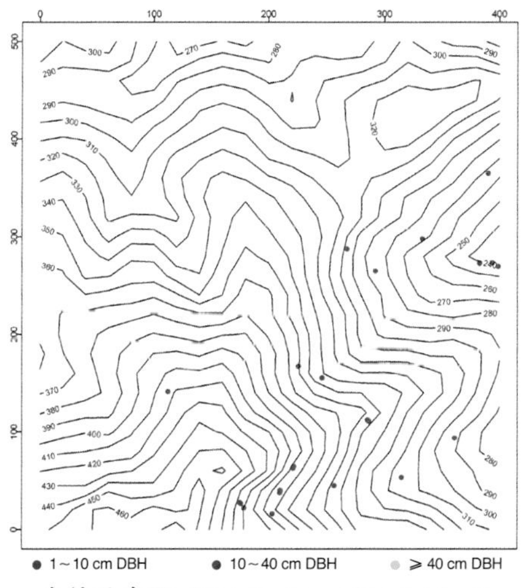

个体分布图 Distribution of individuals

径级分布表 DBH class

胸径等级 (Diameter class) (cm)	个体数 (No. of individuals in the plot)	比例 (Proportion) (%)
1～2	4	16.67
2～5	15	62.50
5～10	5	20.83
10～20	0	0.00
20～30	0	0.00
30～60	0	0.00
≥60	0	0.00

87 虎皮楠 hǔpínán | Oldham Daphniphyllum

Daphniphyllum oldhamii (Hemsl.) Rosenth.
交让木科 | Daphniphyllaceae

代码（SpCode）= DAPOLD
个体数（Individual number/20 hm²）= 1
最大胸径（Max DBH）= 14.1 cm
重要值排序（Importance value rank）= 148

常绿乔木或灌木，高4～15m。小枝纤细，暗黑色。叶纸质，披针形、倒卵状披针形、长圆形或长圆状披针形，边缘反卷，叶背通常被白粉，具细小乳突体。花雌雄异株，花序总状，腋生。核果椭圆或倒卵圆形，具不明显疣状突起，先端具宿存柱头。花期3～5月，果期8～11月。

Evergreen trees or shrubs, 4-15 m tall; branchlets slender, furvous. Leaf blade papery, lanceolate, obovate-lanceolate, oblong, or oblong-lanceolate, margins revolute, abaxially often glaucous, and small papillate. Flowers dioecious, racemes axillary. Drupe ellipsoidal or obovate-globose, tuberculate, calyx absent. Fl. Mar.-May, fr. Aug.-Nov..

叶背 Leaf abaxial surface
摄影：吴林芳 Photo by: Wu Linfang

果序 Infructescence
摄影：吴林芳 Photo by: Wu Linfang

花序 Inflorescence
摄影：吴林芳 Photo by: Wu Linfang

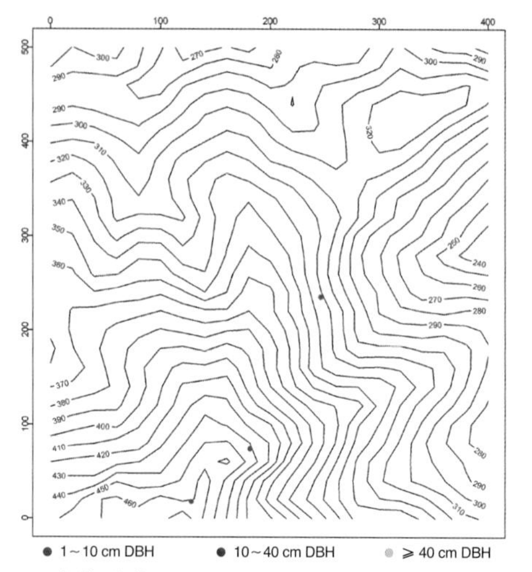

个体分布图 Distribution of individuals

径级分布表 DBH class

胸径等级 (Diameter class) (cm)	个体数 (No. of individuals in the plot)	比例 (Proportion) (%)
1～2	0	0.00
2～5	0	0.00
5～10	0	0.00
10～20	1	100.0
20～30	0	0.00
30～60	0	0.00
≥60	0	0.00

88 小盘木

xiǎopánmù | Microdesmis

Microdesmis caseariifolia Planch. ex. Hook.
小盘木科 | Pandaceae

代码（SpCode）= MICCAS
个体数（Individual number/20 hm^2）= 119
最大胸径（Max DBH）= 17.1 cm
重要值排序（Importance value rank）= 62

常绿灌木或小乔木，高3～8m。小枝初时被毛后脱落。单叶互生，纸质或薄革质，披针形、长圆状披针形或长圆形，边缘具小圆齿或近全缘。花小，黄色，腋生，雄蕊10枚。核果球形，熟时红色，径约5mm，果皮粗糙，2粒种子。花期3～9月，果期7～11月。

Evergreen shrubs or small trees, 3-8 m tall. Young branches pubescent, glabrescent. Leaf blade alternate, lanceolate, oblong-lanceolate, or oblong, papery to thinly leathery, margin crenulate or subentire. Flowers yellow, small, in axillary fascicles, stamens 10. Drupe red when mature, globose, ca. 5 mm, scabrous, 2-seeded. Fl. Mar.-Sep., fr. Jul.-Nov..

核果 Drupe
摄影：曹洪麟 photo by: Cao Honglin

叶 Leaf
摄影：吴林芳 Photo by: Wu Linfang

花枝 Flowering branch
摄影：吴林芳 Photo by: Wu Linfang

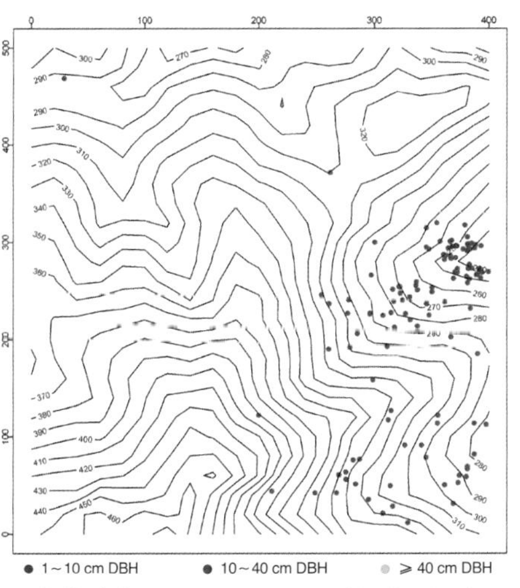

个体分布图 Distribution of individuals

径级分布表 DBH class

胸径等级 (Diameter class) (cm)	个体数 (No. of individuals in the plot)	比例 (Proportion) (%)
1～2	20	16.81
2～5	83	69.75
5～10	15	12.61
10～20	1	0.84
20～30	0	0.00
30～60	0	0.00
≥60	0	0.00

89 鼠刺

shǔcì | Chinese Sweetspire

Itea chinensis Hook. et Arn.

鼠刺科 | Escalloniaceae

代码（SpCode）= ITECHI

个体数（Individual number/20 hm^2）= 77

最大胸径（Max DBH）= 17.5 cm

重要值排序（Importance value rank）= 67

常绿灌木或小乔木，高4～10m。幼枝黄绿色，无毛，老枝褐色，具纵纹。叶薄革质，倒卵形或卵状椭圆形，叶面深绿色，叶背灰绿色，两面无毛。总状花序腋生，单生或2～3朵簇生。蒴果具纵纹，长圆状披针形，6～9mm。花期3～5月，果期5～10月。

Evergreen shrubs or small trees, 4-10 m tall. Young branchlets yellow-green, glabrous, old branchlets brown, striate. Leaf blade abaxially pale green, adaxially deep green, obovate or ovate-elliptic, thinly leathery, both surfaces glabrous. Racemes axillary, solitary or rarely 2- or 3-fascicled. Capsule striate, oblong-lanceolate, 6-9 mm. Fl. Mar.-May, fr. May-Oct..

叶 Leaf

摄影：吴林芳 Photo by: Wu Linfang

果枝 Fruiting branch

摄影：吴林芳 Photo by: Wu Linfang

花枝 Flowring

摄影：吴林芳 Photo by: Wu Linfang

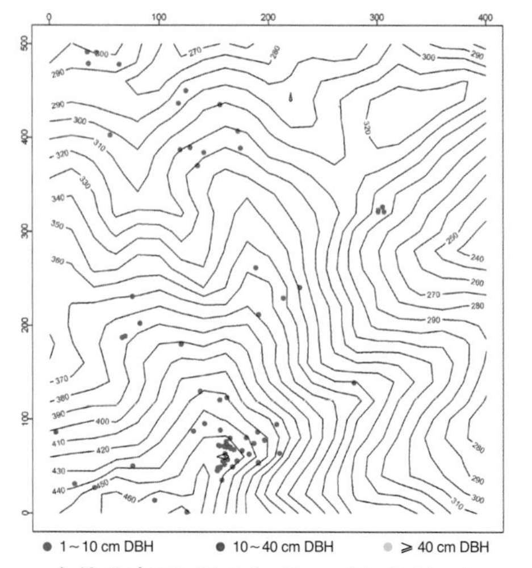

个体分布图 Distribution of individuals

径级分布表 DBH class

胸径等级 (Diameter class) (cm)	个体数 (No. of individuals in the plot)	比例 (Proportion) (%)
1～2	30	38.96
2～5	32	41.56
5～10	12	15.58
10～20	3	3.90
20～30	0	0.00
30～60	0	0.00
≥60	0	0.00

90 香花枇杷（山枇杷） xiānghuāpípá | Wild Loquat

Eriobotrya fragrans Champ. et Benth.
蔷薇科 | Rosaceae

代码（SpCode）= ERIFRA
个体数（Individual number/20 hm^2）= 1
最大胸径（Max DBH）= 1.3 cm
重要值排序（Importance value rank）= 190

常绿小乔木或灌木，高可达10m。小枝粗壮，幼时密被茸毛后脱落。叶革质，长椭圆形，边缘中部以上具疏齿，幼时两面密被短茸毛，不久脱落。圆锥花序顶生，花梗密被棕色茸毛，花瓣基部和子房被毛。梨果球形，被毛，直径1～2.5cm。花期4～5月，果期8～9月。

Evergreen small trees or shrubs, to 10 m tall. Branchlets stout, densely brown tomentose, soon glabrate. Leaf blade oblong-elliptic, leathery, both surfaces densely tomentose when young, glabrescent when old, margin entire basally, remotely inconspicuously serrate apically. Panicle many flowered, peduncle densely brown tomentose, ovary pubescent. Pome globose, 1 2.5 cm in diam.. Fl. Apr.-May, fr. Aug.-Sep..

野外识别特征：

1. 小枝粗壮，幼时密被茸毛，后脱落；
2. 叶革质至厚革质，略聚生枝顶；
3. 叶缘中部以上具疏齿。

Key notes for identification:

1. Branchlets stout, densely brown tomentose, soon glabrate;
2. Leaf blade leathery to thickly leathery, slightly crowded at apex of branchlet;
3. Leaves margin entire basally, remotely inconspicuously serrate apically.

花 Flower
摄影：吴林芳 Photo by: Wu Linfang

果枝 Fruiting branch
摄影：吴林芳 Photo by: Wu Linfang

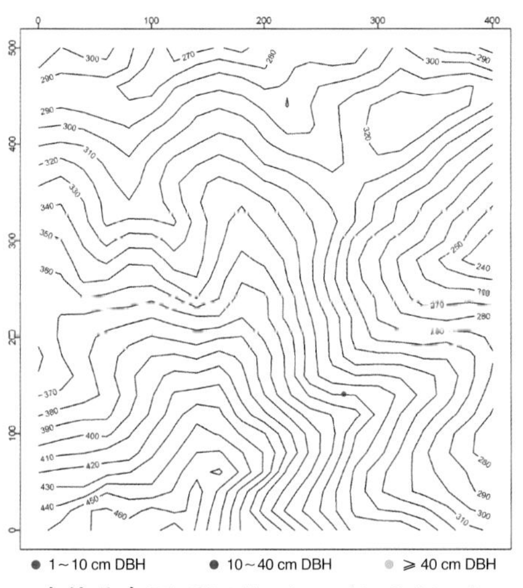

个体分布图 Distribution of individuals

径级分布表 DBH class

胸径等级 (Diameter class) (cm)	个体数 (No. of individuals in the plot)	比例 (Proportion) (%)
1～2	1	100.0
2～5	0	0.00
5～10	0	0.00
10～20	0	0.00
20～30	0	0.00
30～60	0	0.00
≥60	0	0.00

91 腺叶桂樱

xiànyèguìyīng | Wild Cherry

Laurocerasus phaeosticta (Hance) C. K. Schneid
蔷薇科 | Rosacea

代码（SpCode）= LAUPHA
个体数（Individual number/20 hm^2）= 26
最大胸径（Max DBH）= 23.1 cm
重要值排序（Importance value rank）= 91

常绿灌木或小乔木，高4～12m。小枝暗紫褐色，叶背散生黑色小腺点。叶近革质，椭圆形、长圆形至长圆状披针形，两面无毛但网脉明显，全缘，幼苗或萌枝叶具锐齿；叶柄无腺体但叶基具2腺体。总状花序单生于叶腋。核果近球形或横向椭圆形。花期4～5月，果期7～10月。

Evergreen shrubs or small trees, 4-12 m tall. Branchlets dark purplish brown, abaxially scattered black punctuate. Leaf blade elliptic, oblong, oblong-lanceolate, subleathery, both surfaces glabrous, margin entire or on sterile branchlets acutely serrate, petiole without nectaries but leaf base with 2 large flat nectaries near margin. Racemes in axils. Drupe subglobose to transversely ellipsoid. Fl. Apr.-May, fr. Jul.-Oct..

果序 Infructescence
摄影：吴林芳 Photo by: Wu Linfang

叶背 Leaf abaxial surface
摄影：吴林芳 Photo by: Wu Linfang

叶 Leaf
摄影：吴林芳 Photo by: Wu Linfang

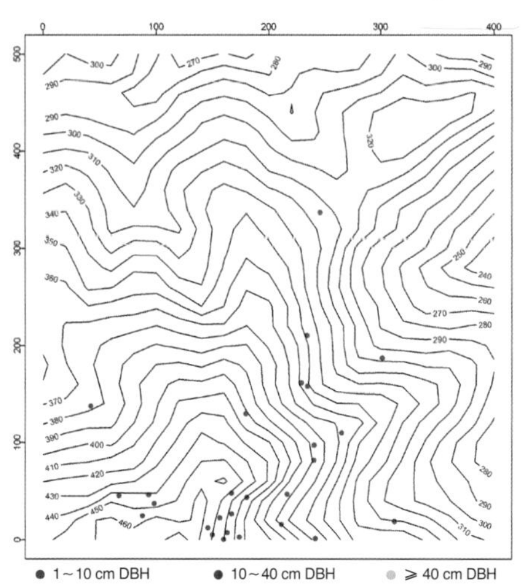

个体分布图 Distribution of individuals

径级分布表 DBH class

胸径等级 (Diameter class) (cm)	个体数 (No. of individuals in the plot)	比例 (Proportion) (%)
1～2	6	23.08
2～5	13	50.0
5～10	6	23.08
10～20	0	0.00
20～30	1	3.85
30～60	0	0.00
≥60	0	0.00

92 桃叶石楠 táoyèshínán | Peachleaf Photinia

Photinia prunifolia (Hooker & Arnott) Lindl.
蔷薇科 | Rosacea

代码（SpCode）= PHOPRU
个体数（Individual number/20 hm^2）= 65
最大胸径（Max DBH）= 25.2 cm
重要值排序（Importance value rank）= 53

常绿乔木，高10～20m。小枝灰黑色，无毛，具皮孔。叶片革质，长圆形或长圆状披针形，两面无毛，叶背布满黑色腺点；叶柄常有锯齿状腺体。复伞房花序顶生，多花，花梗被长茸毛。梨果椭圆形，熟时橙红色。花期3～4月，果期5～12月。

Evergreen trees, 10-20 m tall. Branchlets grayish black, glabrous, with lenticels. Petiole glands dentate; leaf blade oblong or oblong-lanceolate, leathery; both surfaces glabrous; abaxially with black glands. Compound corymbs terminal, many flowered, rachis and pedicels slightly villous. Pome orange when mature, ellipsoid. Fl. Mar.-Apr., fr. May.-Dec..

花序 Inflorescence
摄影：吴林芳 photo by: Wu Linfang

果序 Infructescence
摄影：吴林芳 Photo by: Wu Linfang

叶背 Leaf abaxial surface
摄影：吴林芳 Photo by: Wu Linfang

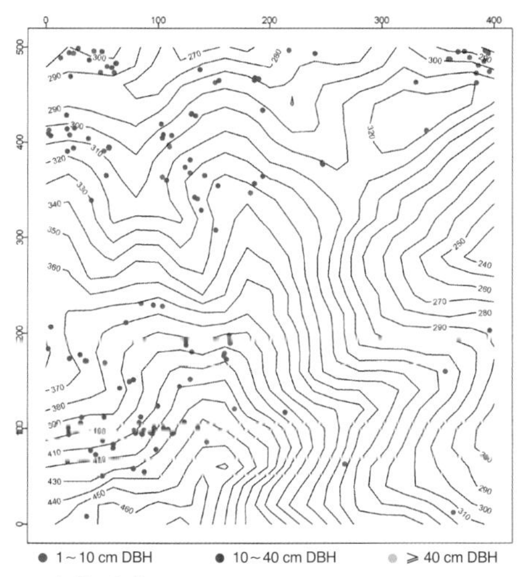

个体分布图 Distribution of individuals

径级分布表 DBH class

胸径等级 (Diameter class) (cm)	个体数 (No. of individuals in the plot)	比例 (Proportion) (%)
1～2	0	0.00
2～5	9	13.85
5～10	12	18.46
10～20	42	64.62
20～30	2	3.08
30～60	0	0.00
≥60	0	0.00

93 臀果木 túnguǒmù | Pygeum

Pygeum topengii Merr.
蔷薇科 | Rosacea

代码（SpCode）= PYGTOP
个体数（Individual number/20 hm²）= 108
最大胸径（Max DBH）= 58.8 cm
重要值排序（Importance value rank）= 55

常绿乔木，高25m。小枝具皮孔，幼时被毛，老时无毛。叶薄革质或纸质，卵状椭圆形或椭圆形，近基部有2黑色腺体，全缘。总状花序有花多数，单生或2至数枝簇生于叶腋。核果肾形，无毛，深褐色。花期6~9月，果期冬季。

Evergreen trees, 25 m tall. Branchlets brown pubescent when young, glabrescent, with orbicular small lenticels. Leaf blade ovate-elliptic or elliptic, thinly leathery or papery, with 2 black nectaries near base, margin entire. Racemes solitary or to several in a fascicle, many-flowered. Drupe dark brown, reniform, glabrous. Fl. Jun.-Sep., fr. winter.

果序 Infructescence
摄影：吴林芳 Photo by: Wu Linfang

叶 Leaf
摄影：吴林芳 Photo by: Wu Linfang

花序 Inflorescence
摄影：吴林芳 Photo by: Wu Linfang

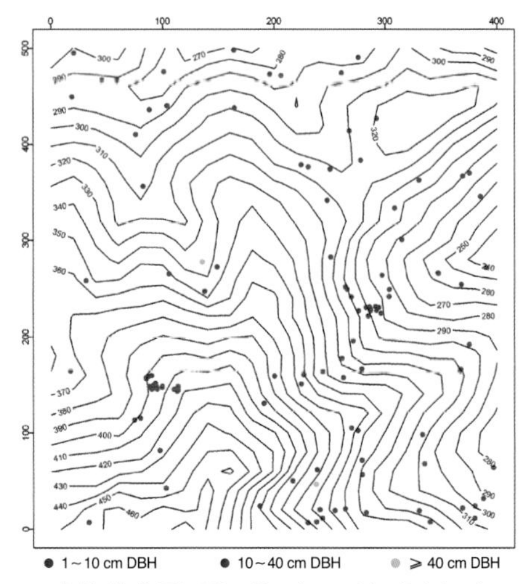

个体分布图 Distribution of individuals

径级分布表 DBH class

胸径等级 (Diameter class) (cm)	个体数 (No. of individuals in the plot)	比例 (Proportion) (%)
1~2	35	32.41
2~5	25	23.15
5~10	7	6.48
10~20	19	17.59
20~30	14	14.96
30~60	8	7.41
≥60	0	0.00

94 石斑木（春花） shíbānmù I Indic Raphiolepis

Rhaphiolepis indica (Linn.) Lindl.
蔷薇科 I Rosacea

代码（SpCode）= RAPIND
个体数（Individual number/20 hm^2）= 3
最大胸径（Max DBH）= 2 cm
重要值排序（Importance value rank）= 183

常绿灌木或小乔木，高1～4m。叶常聚生于枝顶，叶革质，长圆形至卵状披针形，边缘具疏而不齐的钝齿；叶柄长5～18mm，无毛。圆锥花序顶生，多或少花；花白色或粉红色。果球形，紫黑色。花期2～4月，果期5～12月。

Evergreen shrubs or small trees, 1-4 m tall. Leaves often crowded at apex of branchlet; petioles 5-18 mm, glabrous; leaf blade leathery, oblong to ovate-lanceolate, margin dentate. Panicles or racemes terminal, many-flowered; petals white or pink. Pome globose, purplish blak. Fl. Feb.-Apr., fr. May-Dec..

花序 Inflorescence
摄影：吴林芳 Photo by: Wu Linfang

果枝 Fruiting branch
摄影：吴林芳 Photo by: Wu Linfang

叶 Leaf
摄影：吴林芳 Photo by: Wu Linfang

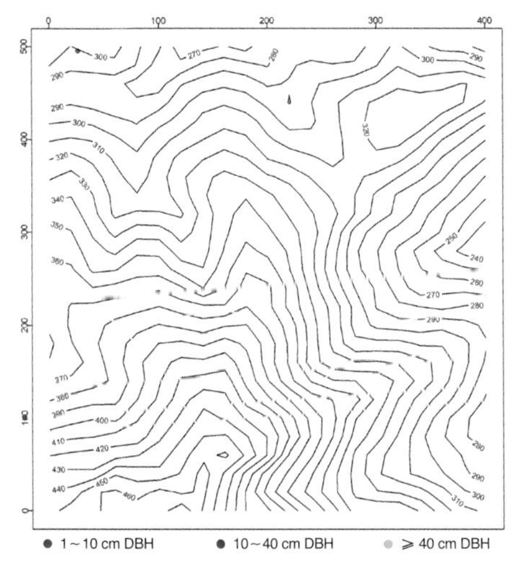

个体分布图 Distribution of individuals

径级分布表 DBH class

胸径等级 (Diameter class) (cm)	个体数 (No. of individuals in the plot)	比例 (Proportion) (%)
1～2	3	100.00
2～5	0	0.00
5～10	0	0.00
10～20	0	0.00
20～30	0	0.00
30～60	0	0.00
≥60	0	0.00

95 海红豆

hǎihóngdòu | Red Sandalwood

Adenanthera microsperma Teijsm. & Binned., Natuurk.
含羞草科 | Mimosaceae

代码（SpCode）= ADEMIC
个体数（Individual number/20 hm^2）= 2
最大胸径（Max DBH）= 27.6 cm
重要值排序（Importance value rank）= 149

落叶乔木，高5～20m。小枝被毛。二回羽状复叶，叶柄和叶轴均被毛，并有腺体；羽片3～5对；小叶4～7对互生，长圆形或卵形，两面被毛，具短柄。总状花序单生于叶腋或在枝顶排成圆锥花序。荚果狭长圆形。花4～7月，果7～10月。

Deciduous trees, 5-20 m tall. Branchlets puberulent. Leaves bipinnate. Petioles and leaf-rachis puberulent, eglandular. Pinnae 3-5 pairs. Leaflets 4-7 pairs, alternate, oblong or ovate. Puberulent on both surfaces, with short petiolules. Racemes axillary, simple or forming terminal panicle on branchlent. Pods narrowly oblong. Fr. Apr.-Jul., fr. Jul.-Oct..

野外识别特征：

1. 叶二回羽状复叶，羽片3～5对，近对生；小叶4～7对，互生；
2. 小叶纸质，被毛，具短柄，较小；
3. 小叶中脉在中间。

Key notes for identification:

1. Leaves bipinnate, pinnae 3-5 pairs, subopposition; leaflets 4-7 pairs, alternate.
2. Leaflets papery, puberulent on both surfaces, with short petiolules, relatively small.
3. Leaflets midrib in middle.

花序 Inflorescence
摄影：吴林芳 Photo by: Wu Linfang

果枝 Fruiting branch
摄影：吴林芳 Photo by: Wu Linfang

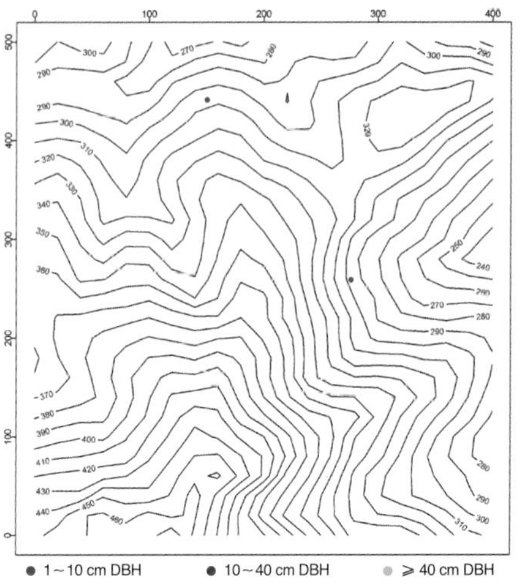

个体分布图 Distribution of individuals

径级分布表 DBH class

胸径等级 (Diameter class) (cm)	个体数 (No. of individuals in the plot)	比例 (Proportion) (%)
1～2	0	0.00
2～5	0	0.00
5～10	0	0.00
10～20	1	50.0
20～30	1	50.0
30～60	0	0.00
≥60	0	0.00

96 猴耳环 hóuěrhuán | Monkeypod

Archidendron clypearia (Jack.) Nielsen
含羞草科 | Mimosaceae

代码（SpCode）= ARCCLY
个体数（Individual number/20 hm^2）= 10
最大胸径（Max DBH）= 2.7 cm
重要值排序（Importance value rank）= 111

常绿乔木，高10m。小枝明显有棱，密被黄褐色茸毛。二回羽状复叶，羽片3～8对，通常4～5对；小叶革质，斜菱形，两面被毛。圆锥花序顶生或腋生。荚果旋卷，边缘在种子间缢缩。花期2～6月，果期4～8月。

Evergreen trees, 10 m tall. Branchelets acutely angulate, densely yellow-tomentose. Leaves bipinnate, pinnae (3-) 4-5 (-8) pairs, leaflets leathery, oblique rhombic, slightly brown pubescent on both surfaces. Heads of several flowers forming terminal or axillary panicle. Pods twisted, indented between the seeds on the outer edge. Fl. Feb.-Jun., fr. Apr.-Aug..

野外识别特征：
1. 小枝具棱，密被黄褐色茸毛；
2. 小叶片革质，斜菱形，两面被毛；
3. 二回羽状复叶，羽片通常4～5对，基部小叶3～6对，顶部小叶10～12对。

Key notes for identification:
1. Branchlets angulate, densely tomentose.
2. Leaflets leathery, oblique rhombic, pubescent on both surfaces.
3. Leaves bipinnate, pinnae usually 4-5 pairs, lowermost leaflets 3-6 pairs, uppermost leaflets 10-12 pairs.

花序 Inflorescence
摄影：吴林芳 Photo by: Wu Linfang

果枝 Fruiting branch
摄影：吴林芳 Photo by: Wu Linfang

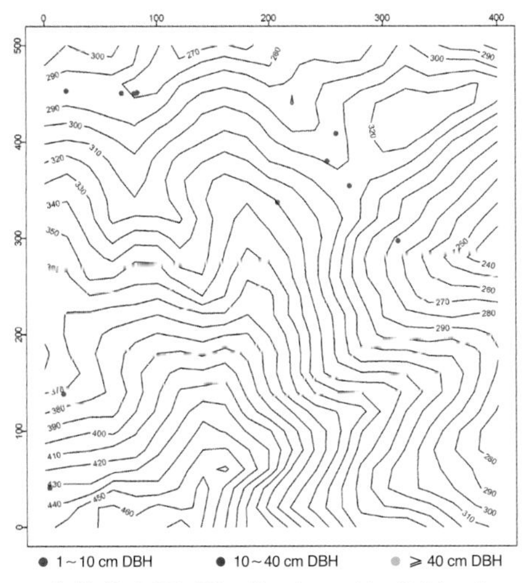

个体分布图 Distribution of individuals

径级分布表 DBH class

胸径等级 (Diameter class) (cm)	个体数 (No. of individuals in the plot)	比例 (Proportion) (%)
1～2	4	40.00
2～5	6	60.00
5～10	0	0.00
10～20	0	0.00
20～30	0	0.00
30～60	0	0.00
≥60	0	0.00

97 亮叶猴耳环

liàngyèhóuěrhuán | Chinese Apea Earring

Archidendron lucidum (Benth.) Nielsen
含羞草科 | Mimosaceae

代码（SpCode）= ARCLUC
个体数（Individual number/20 hm²）= 266
最大胸径（Max DBH）= 17.3 cm
重要值排序（Importance value rank）= 39

常绿乔木，高2～10m。小枝、叶柄和花序被褐色短茸毛。羽片1～2对，叶柄基部和小叶片叶轴有圆形而凹陷的腺体。小叶在羽下部常2～3对，上部常4～5对，小叶片斜卵形或长圆形。头状花序球形，10～20朵花排成顶生的圆锥花序。荚果旋卷成圈。花期4～6月，果期7～12月。

Evergreen trees, 2-10 m tall. Branchlets, petioles and inflorescence shortly brown-tomentose. Pinnae 1-2 pairs base of petioles, and leaf rachis with round sunken glands, leaflets 2-3 pairs in lower pinnae, 4-5 pairs in upper pinnae, obliquely ovate or oblong. Heads globular, with 10-20 flowers, forming a long terminal panicle. Pods twisted into a ring. Fl. Apr.-Jun., fr. Jul.-Dec..

荚果 Legume
摄影：吴林芳 Photo by: Wu Linfang

果枝 Fruiting branch
摄影：吴林芳 Photo by: Wu Linfang

花枝 Flowering branch
摄影：吴林芳 Photo by: Wu Linfang

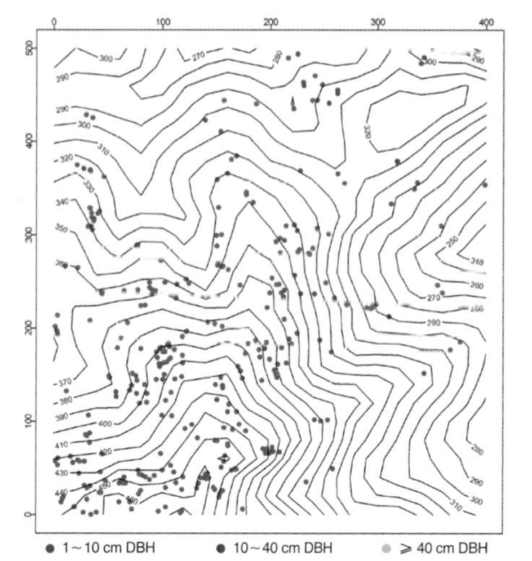

个体分布图 Distribution of individuals

径级分布表 DBH class

胸径等级 (Diameter class) (cm)	个体数 (No. of individuals in the plot)	比例 (Proportion) (%)
1～2	132	49.62
2～5	116	43.61
5～10	13	4.89
10～20	5	1.88
20～30	0	0.00
30～60	0	0.00
≥60	0	0.00

102 软荚红豆

ruǎnjiáhóngdòu | Soft-fruited Ormosia

Ormosia semicastrata Hance
蝶形花科 | Papilionaceae

代码（SpCode）= ORMSEM
个体数（Individual number/20 hm^2）= 33
最大胸径（Max DBH）= 14.8 cm
重要值排序（Importance value rank）= 86

常绿乔木，高达12m。树皮褐色至灰白色。小枝疏被黄褐色柔毛。奇数羽状复叶，叶轴于顶部小叶对外延伸12～20mm；有小叶3～9片，革质，卵状长椭圆形至椭圆形。荚果小，近圆形，略肿胀，光亮。种子1，红色。花期3～5月，果期5～12月。

Evergreen trees, to 12 m tall. Bark brown or grey. Branchlets sparsely tawny pubescent. Leaves odd-pinnate, rachis elongating from the apical pair of leaflets to 12-20 mm. Leaflets 3-9, leathery, ovate-elliptic to elliptic. Pods suborbicular, slightly turgid, glabrous. Seed 1, scarlet. Fl. Mar.-May, fr. May-Dec..

野外识别特征：

1. 树皮褐色至灰白色，具皮孔；
2. 小叶革质，两面无毛；
3. 荚果小，近圆形，略肿胀，无毛，种子1。

Key notes for identification:

1. Bark brown or grey, lenticellate.
2. Leaflets leathery, glabrous on both surfaces.
3. Pods suborbicular, slightly small, slightly turgid, glabrous, seed 1.

叶背 Leaf abaxial surface
摄影：吴林芳 Photo by: Wu Linfang

叶 Leaf
摄影：吴林芳 Photo by: Wu Linfang

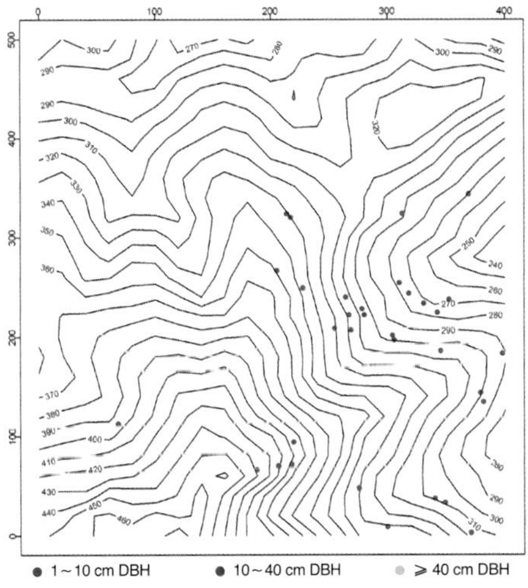

个体分布图 Distribution of individuals

径级分布表 DBH class

胸径等级 (Diameter class) (cm)	个体数 (No. of individuals in the plot)	比例 (Proportion) (%)
1～2	13	39.39
2～5	10	30.30
5～10	4	12.12
10～20	6	18.18
20～30	0	0.00
30～60	0	0.00
≥60	0	0.00

103 锥（锥栗） zhuī | Chinese Chestnut

Castanopsis chinensis (Sprengel) Hance
壳斗科 | Fagaceae

代码（SpCode）= CASCHI
个体数（Individual number/20 hm^2）= 2544
最大胸径（Max DBH）= 175.0 cm
重要值排序（Importance value rank）= 1

常绿大乔木，高达25m。树皮纵裂，片状脱落。枝叶均无毛。单叶互生，稍二列，革质，披针形，中部以上有锐齿，两面同色。每壳斗常有雌花1。壳斗圆球形，连刺径25～35mm；坚果圆锥形，12～16mm×10～13mm。花5～7月，果翌年9～11月熟。

Great evergreen trees, to 25 m tall. Bark slit, flaky break off. Branches and leaf blades glabrous. Leaves alternate, slightly distichous. Leaf blade lanceolate, leathery, margin at least from middle to apex serrate, concolorous. Female flowers 1 per cupule. Cupule globose, 25-35 mm in diam.. Nut conical, 1.2-1.6 cm × 1-1.3 cm. Fl. May-Jul., fr. Sep.-Nov. of next year.

树干 Trunk
摄影：吴林芳 Photo by: Wu Linfang

花枝 Flowering branch
摄影：吴林芳 Photo by: Wu Linfang

果枝及坚果 Fruiting branch & nut
摄影：吴林芳 Photo by: Wu Linfang

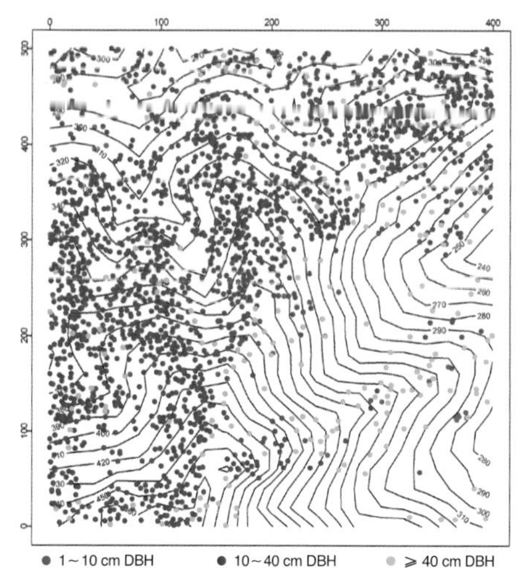

个体分布图 Distribution of individuals

径级分布表 DBH class

胸径等级 (Diameter class) (cm)	个体数 (No. of individuals in the plot)	比例 (Proportion) (%)
1～2	82	3.22
2～5	96	3.77
5～10	97	3.81
10～20	526	20.68
20～30	877	34.47
30～60	764	30.03
≥60	102	4.01

104 黧蒴锥 líshuòzhuī | Fissure Chestnut

Castanopsis fissa (Champ. ex Bentham) R. et W.
壳斗科 | Fagaceae

代码（SpCode）= CASFIS
个体数（Individual number/20 hm^2）= 248
最大胸径（Max DBH）= 31.3 cm
重要值排序（Importance value rank）= 50

常绿乔木，高达20m。先锋速生树种。叶互生，螺旋状着生，厚纸质，矩圆形至倒卵状椭圆形，较大，常15～25cm × 5～9cm；叶缘基部到中部常波浪状并有圆齿。壳斗和坚果均圆球形或宽椭圆形，坚果顶部被棕红色毛。花期4～6月，果期10～12月。

Evergreen trees, to 20 m tall. A pioneer fast-growing species. Leaves alternate, slightly spirally arranged. Leaf blade oblong to obovate-elliptic, often 15-25 cm × 5-9 cm, thickly papery, margin from base to middle undulate and crenate. Cupule and nut globose to elliptic, nut apically brown-red tomentose. Fl. Apr.-Jun., fr. Oct.-Dec..

叶 Leaf
摄影：吴林芳 photo by: Wu Linfang

果序 Infructescence
摄影：吴林芳 Photo by: Wu Linfang

花序 Inflorescence
摄影：吴林芳 Photo by: Wu Linfang

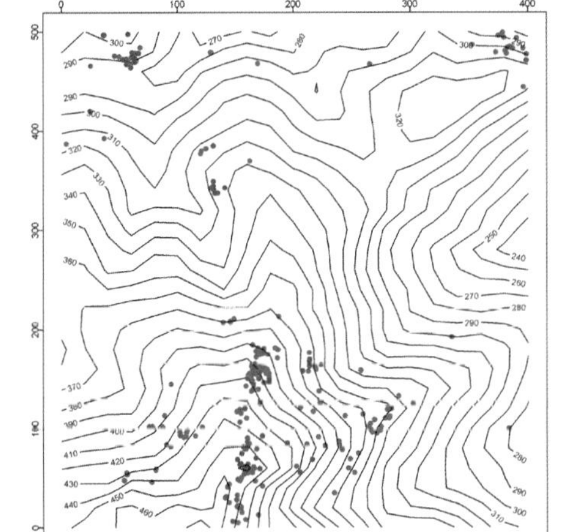

个体分布图 Distribution of individuals

径级分布表 DBH class

胸径等级 (Diameter class) (cm)	个体数 (No. of individuals in the plot)	比例 (Proportion) (%)
1～2	106	42.74
2～5	73	29.44
5～10	30	12.10
10～20	23	9.27
20～30	15	6.05
30～60	1	0.40
≥60	0	0.00

105 白颜树

báiyánshù | Subaequal Gironniera

Gironniera subaequalis Planch.
榆科 | Ulmaceae

代码（SpCode）= GIRSUB
个体数（Individual number/20 hm^2）= 290
最大胸径（Max DBH）= 56.3 cm
重要值排序（Importance value rank）= 34

常绿乔木，高10～20（～30）m。雌雄异株。树皮灰至深灰色，较平滑。小枝常“之”字型，淡黄绿色，被毛。叶革质，椭圆形或椭圆状矩圆形，叶面亮绿无毛，叶背绿色稍粗糙，叶脉被毛，近全缘，仅在顶部有疏齿。核果大体卵形或椭圆形，径4～8mm，果皮熟时橙红色。花期2～4月，果期几全年。

Evergreen trees, 10-20 (-30) m tall. Dioecious. Bark gray to dark gray, smooth. Branchlets usually zig-zag, yellowish green, covered with hirsute hairs. Leaf blade elliptic to elliptic-oblong, leathery, abaxially green, scabridulous, and with appressed hairs on major veins, adaxially green and glabrous; margin bluntly serrate or subentire. Drupes ovoid to ellipsoid, 4-8 mm in diam., endocarp reddish orange when mature, Fl. Feb.-Apr., fr. most of the year.

果枝 Fruiting branch
摄影：吴林芳 Photo by: Wu Linfang

枝叶 Branch & leaves
摄影：吴林芳 Photo by: Wu Linfang

花枝 Flowering branch
摄影：吴林芳 Photo by: Wu Linfang

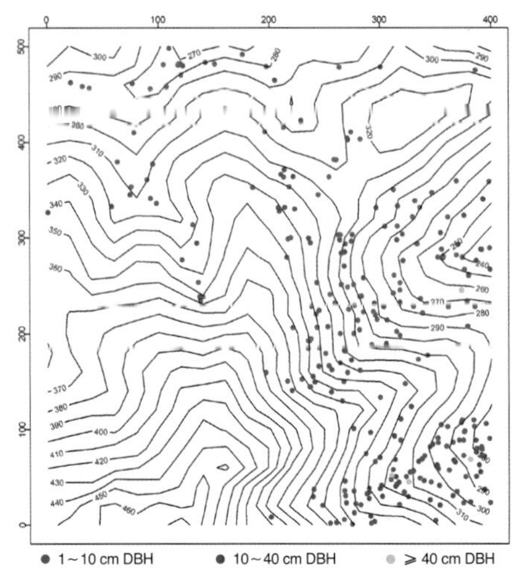

个体分布图 Distribution of individuals

径级分布表 DBH class

胸径等级 (Diameter class) (cm)	个体数 (No. of individuals in the plot)	比例 (Proportion) (%)
1～2	18	6.21
2～5	77	26.55
5～10	68	23.45
10～20	81	27.93
20～30	30	10.34
30～60	16	5.52
≥60	0	0.00

106 狭叶山黄麻

xiáyèshānhuángmá | Narrowleaf Trema

Trema angustifolia (Planch.) Bl.
榆科 | Ulmaceae

代码（SpCode）= TREANG
个体数（Individual number/20 hm^2）= 3
最大胸径（Max DBH）= 4.6 cm
重要值排序（Importance value rank）= 157

常绿灌木或小乔木。小枝纤细，紫红色，密被细粗毛。叶互生，纸质，卵状披针形，边缘细锯齿，叶粗糙，基出脉3条。花单性，雌雄异株或同株。核果宽卵状或近圆球形，1.5～2.5mm，橘红色，有宿存的花被。花期4～6月，果期8～11月。

Evergreen shrubs or small trees. Branchlets slender, reddish purple, densely hirsute. Leaves alternate, leaf blade lanceolate to narrowly lanceolate, margin denticulate, scabrous, basally 3-veined. Flowers unisexual. Drupes reddish orange when mature, drupes ovoid to globose 1.5-2.5 mm; perianth persistent. Fl. Apr.-Jun., fr. Aug.-Nov..

野外识别特征：

1. 小枝纤细，紫红色；
2. 叶卵状披针形，叶柄2～5mm；
3. 果熟时橘红色。

Key notes for identification:

1. Branchlets slender, reddish purple.
2. Leaf blade lanceolate to narrowly lanceolate, petiole 2-5 mm.
3. Drupes reddish orange when mature.

叶 Leaf
摄影：吴林芳 Photo by: Wu Linfang

叶背 Leaf abaxial surface
摄影：吴林芳 Photo by: Wu Linfang

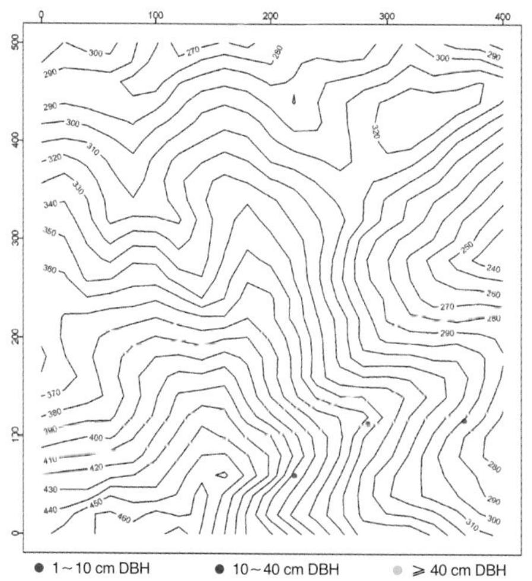

个体分布图 Distribution of individuals

径级分布表 DBH class

胸径等级 (Diameter class) (cm)	个体数 (No. of individuals in the plot)	比例 (Proportion) (%)
1～2	1	33.33
2～5	2	66.67
5～10	0	0.00
10～20	0	0.00
20～30	0	0.00
30～60	0	0.00
≥60	0	0.00

107 山黄麻

shānhuángmá | India-charcoal Trema

Trema tomentosa (Roxb.) Hara

榆科 | Ulmaceae

代码（SpCode）= TRETOM

个体数（Individual number/20 hm^2）= 7

最大胸径（Max DBH）= 3.6 cm

重要值排序（Importance value rank）= 128

常绿小乔木或灌木，高10m。树皮灰褐色，平滑或细龟裂；小枝灰褐色至棕褐色，被毛。叶纸质或薄革质，宽卵形或卵状矩圆形，7～15（～20）cm×3～7（～8）cm，基出脉3，边缘有细锯齿，叶柄长7～18mm。核果熟时紫黑色，压扁，具宿存的花被。花期3～6月，果期9～11月。

Evergreen small trees or shrubs, to 10 m tall. Bark grayish brown, smooth or fissured. Branchlets grayish brown to brown, densely pubescent. Petiole 0.7-1.8 cm. Leaf blade 7-15(-20) cm × 3-7(-8) cm, ovate or ovate-oblong, papery or thinly leathery, margin denticulate, basally 3-veined. Drupes blackish purple when mature, compressed, perianth persistent. Fl. Mar.-Jun., fr. Sep.-Nov..

野外识别特征：

1. 小枝灰褐色至棕褐色；
2. 叶宽卵形或卵状矩圆形，叶柄7～18mm；
3. 果熟时紫黑色。

Key notes for identification:

1. Branchlets grayish brown to brown.
2. Leaf blade ovate or ovate-oblong, petiole 7-18 mm.
3. Drupes blackish purple when mature.

果枝 Fruiting branch

摄影：吴林芳 Photo by: Wu Linfang

叶 Leaf

摄影：吴林芳 Photo by: Wu Linfang

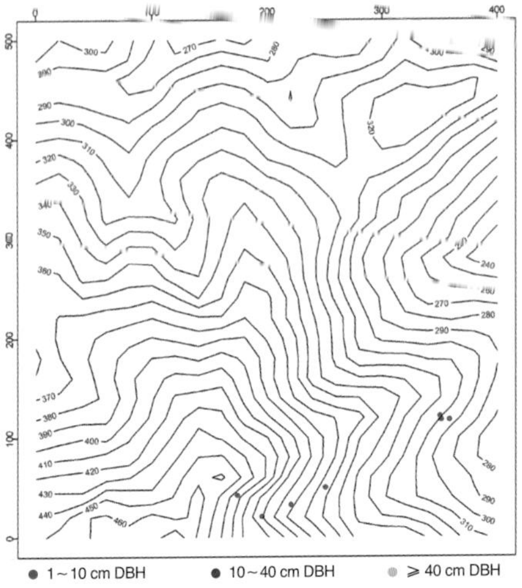

个体分布图 Distribution of individuals

径级分布表 DBH class

胸径等级 (Diameter class) (cm)	个体数 (No. of individuals in the plot)	比例 (Proportion) (%)
1～2	1	14.29
2～5	6	85.71
5～10	0	0.00
10～20	0	0.00
20～30	0	0.00
30～60	0	0.00
≥60	0	0.00

108 二色波罗蜜（小叶胭脂） èrsèbōluómì | Bicolor Artocarpus

Artocarpus styracifolius Pierre
桑科 | Moraceae

代码（SpCode）= ARTSTY
个体数（Individual number/20 hm^2）= 417
最大胸径（Max DBH）= 34.6 cm
重要值排序（Importance value rank）= 29

常绿乔木，高达20m。雌雄同株。树皮暗灰色，粗糙。小枝幼时被白色短柔毛。叶二列，革质到纸质，长圆形、倒卵状披针形，4～8cm×2.5～3cm，幼枝的叶常分裂或在上部有锯齿，叶背被毛。聚花果球形，黄色，径约4cm，被毛；核果球形。花早秋，果秋末冬初。

Evergreen trees to 20 m tall. Monoecious. Bark dark gray, rough. Young branchlets densely white appressed-puberulent. Leaves distichous. Leaf blade oblong, obovate-lanceolate, 4-8 cm × 2.5-3 cm, lobed or apically shallowly toothed on new leaves of young trees, abaxially densely covered white farinaceous hairs, leathery to papery. Fruiting syncarp yellow, globose, ca. 4 cm in diam., pubescent, drupes globose. Fl. early autumn, fr. late autumn to early winter.

野外识别特征：
1. 叶二列；
2. 幼枝的叶常分裂或在上部有锯齿；
3. 叶4～8cm×2.5～3cm，先端尾渐尖。

Key notes for identification:
1. Leaves distichous.
2. Leaf blade lobed or apically shallowly toothed on new leaves of young trees.
3. Leaf blade 4-8 cm × 2.5-3 cm, apex acuminate to caudate.

果 Fruit
摄影：吴林芳 Photo by: Wu Linfang

叶及花序 Leaf & Inflorescence
摄影：吴林芳 Photo by: Wu Linfang

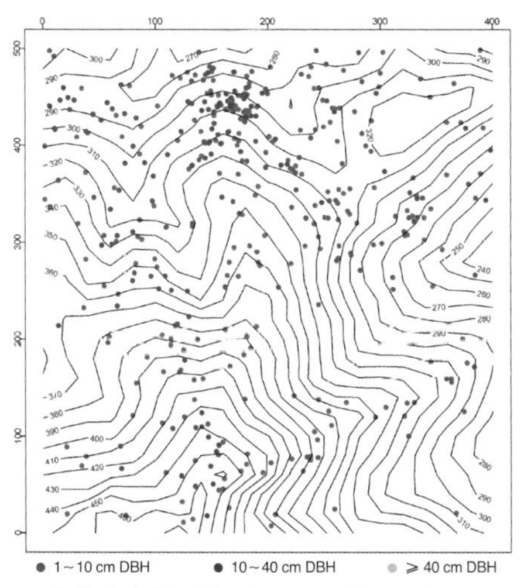

个体分布图 Distribution of individuals

径级分布表 DBH class

胸径等级 (Diameter class) (cm)	个体数 (No. of individuals in the plot)	比例 (Proportion) (%)
1～2	72	17.27
2～5	116	27.82
5～10	89	21.34
10～20	94	22.54
20～30	37	8.87
30～60	9	2.16
≥60	0	0.00

109 胭脂 yānzhī | Tonkin Artocarpus

Artocarpus tonkinensis A. Chev. ex Gagnep.
桑科 | Moraceae

代码（SpCode）= ARTTON
个体数（Individual number/20 hm²）= 2
最大胸径（Max DBH）= 20.3 cm
重要值排序（Importance value rank）= 154

常绿乔木，高16m。雌雄同株。树皮褐色，粗糙。叶二列，革质，椭圆形、长圆形或倒卵形，8～20cm×4～10cm；全缘，有时先端有浅齿；叶背被毛。花序单生叶腋。聚花果近球形，黄色，径约6.5cm；核果椭圆形。花期夏秋，果期冬季。

Evergreen trees to 16 m tall. Monoecious. Bark brown, rough. Leaves distichous. Leaf blade elliptic, obovate, or oblong, 8-20 cm × 4-10 cm, margin entire or sometimes apically with a few shallow teeth, abaxially densely pubescent, leathery. Inflorescences axillary, solitary. Fruiting syncarp yellow, subglobose, ca. 6.5 cm in diam.. Drupes ellipsoid. Fl. summer to autumn, fr. winter.

野外识别特征：
1. 叶二列；
2. 幼叶不变型；
3. 叶8～20 cm × 4～10 cm，先端短尖。

Key notes for identification:
1. Leaves distichous.
2. Leaves no deformed when young.
3. Leaf blade 8-20 cm × 4-10 cm, apex mucronate.

叶 Leaf
摄影：吴林芳 Photo by: Wu Linfang

叶背 Leaf abaxial surface
摄影：吴林芳 Photo by: Wu Linfang

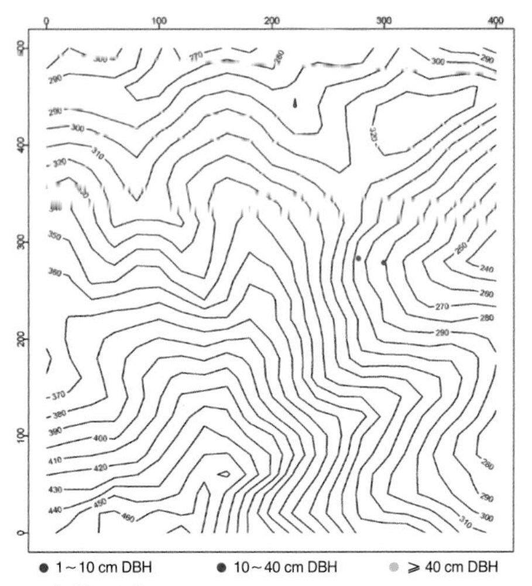

个体分布图 Distribution of individuals

径级分布表 DBH class

胸径等级 (Diameter class) (cm)	个体数 (No. of individuals in the plot)	比例 (Proportion) (%)
1～2	0	0.00
2～5	0	0.00
5～10	0	0.00
10～20	1	50.00
20～30	1	50.00
30～60	0	0.00
≥60	0	0.00

110 黄毛榕 huángmáoróng | Fulvous Fig

Ficus esquiroliana H. Lévl.
桑科 | Moraceae

代码（SpCode）= FICESQ
个体数（Individual number/20 hm^2）= 53
最大胸径（Max DBH）= 34.0 cm
重要值排序（Importance value rank）= 77

常绿小乔木或灌木，高4～10m。树皮灰褐色，具纵棱。幼枝常中空，被褐黄色硬长毛。叶互生，纸质，广卵形，长17～27cm，宽12～20cm；叶背被褐黄色长毛，边缘有细锯齿。榕果腋生，圆锥状椭圆形，径2～2.5cm，密被浅褐色长毛。花果期5～7月。

Evergreen small trees or shrubs, 4-10 m tall. Bark grayish brown, with longitudinal ridges. Twigs often empty in middle. Stiffly brownish yellow hirsute. Leaves alternate. Leaf blade broadly obovate, 17 27 cm × 12-20 cm, papery, abaxially with brownish yellow soft felted hairs. margin sparsely serrate. Figs axillary, solitary, ovoid, 2-2.5 cm in diam., pale brown hirsute. Fl. and fr. May-Jul..

叶背 Leaf abaxial surface
摄影：吴林芳 Photo by: Wu Linfang

果枝 Fruiting branch
摄影：吴林芳 Photo by: Wu Linfang

叶 Leaf
摄影：吴林芳 Photo by: Wu Linfang

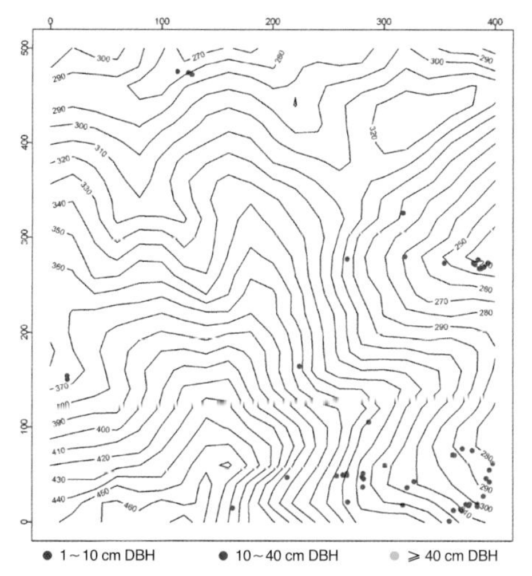

个体分布图 Distribution of individuals

径级分布表 DBH class

胸径等级 (Diameter class) (cm)	个体数 (No. of individuals in the plot)	比例 (Proportion) (%)
1～2	5	9.43
2～5	28	52.83
5～10	17	32.08
10～20	2	3.77
20～30	0	0.00
30～60	1	1.89
≥60	0	0.00

111 水同木

shuǐtóngmù | Common Yellow Stem Fig

Ficus fistulosa Reinw. ex Bl.
桑科 | Moraceae

代码（SpCode）= FICFIS
个体数（Individual number/20 hm²）= 13
最大胸径（Max DBH）= 8.9 cm
重要值排序（Importance value rank）= 110

常绿小乔木。雌雄异株。树皮黑褐色，枝被硬毛。叶互生，纸质，倒卵形至长圆形，长10～20cm，宽4～7cm，全缘或微波状，叶面无毛，叶背被疏毛或黄色小突体。榕果簇生于老干发出的瘤状枝上，橙红色，近球形，径1.5～2cm。花果期5～7月。

Evergreen small trees, dioecious. Bark dark brown, branchlets hispid. Leaves alternate. Leaf blade obovate to oblong, 10-20 cm × 4-7 cm, papery, abaxially sparsely pubescent or yellow tuberculate, adaxially glabrous. Figs on short conic branchlets on main branches, reddish orange when mature, subglobose, 1.5-2 cm, subglabrous. Fl. May.-Jul..

果枝 Fruiting branch
摄影：吴林芳 Photo by: Wu Linfang

叶 Leaf
摄影：吴林芳 Photo by: Wu Linfang

叶背 Leaf abaxial surface
摄影：吴林芳 Photo by: Wu Linfang

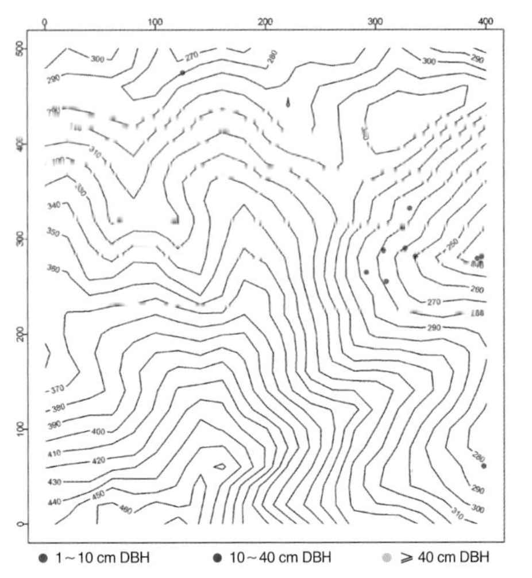

个体分布图 Distribution of individuals

径级分布表 DBH class

胸径等级 (Diameter class) (cm)	个体数 (No. of individuals in the plot)	比例 (Proportion) (%)
1～2	5	38.46
2～5	7	53.85
5～10	1	7.69
10～20	0	0.00
20～30	0	0.00
30～60	0	0.00
≥60	0	0.00

112 粗叶榕 cūyèróng | Hairy Fig

Ficus hirta Vahl.
桑科 | Moraceae

代码（SpCode）= FICHIR
个体数（Individual number/20 hm^2）= 3
最大胸径（Max DBH）= 3.2 cm
重要值排序（Importance value rank）= 146

常绿灌木或小乔木。嫩枝常中空。小枝、叶和榕果被金黄色开展的长硬毛。叶互生，纸质，多型，边缘具细齿，有时全缘或3～5深裂。榕果成对腋生，球形或椭圆球形，无梗或近无梗，径1～3 (3.5) cm。花果几全年。

Evergreen shrubs or small trees. Twigs often empty in middle. Branchlets, leaves and figs golden yellow hirsute. Leaves alternate. Leaf blade simple or palmately 3-5-lobed, papery, margin entire or with small serrations. Figs axillary on normal leafy shoots, paired, globose or ellipsoid, 1-3 (-3.5) cm in diam., sessile. Fl. and fr. almost whole year.

榕果 Multlple fruit
摄影：吴林芳 photo by: Wu Linfang

果枝 Fruiting branch
摄影：吴林芳 Photo by: Wu Linfang

叶 Leaf
摄影：吴林芳 Photo by: Wu Linfang

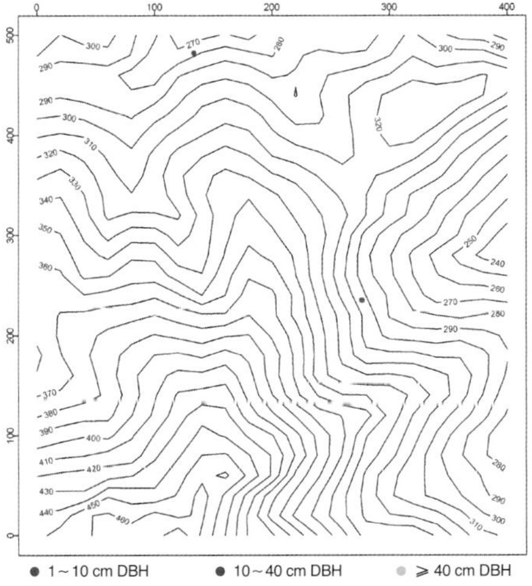

个体分布图 Distribution of individuals

径级分布表 DBH class

胸径等级 (Diameter class) (cm)	个体数 (No. of individuals in the plot)	比例 (Proportion) (%)
1～2	1	33.3
2～5	2	66.67
5～10	0	0.00
10～20	0	0.00
20～30	0	0.00
30～60	0	0.00
≥60	0	0.00

113 九丁榕（凸脉榕） jiǔdīngróng | Veined Fig

Ficus nervosa Heyne ex Roth.
桑科 | Moraceae

代码（SpCode）= FICNER
个体数（Individual number/20 hm²）= 6
最大胸径（Max DBH）= 26.6 cm
重要值排序（Importance value rank）= 115

常绿乔木。树皮深褐色。小枝被短毛。叶革质，倒披针形，长圆形至椭圆形，长6～15（～20）cm，宽2.5～6cm，边缘全缘，中脉和侧脉在叶背明显突起。榕果黄色或淡红棕色，单生或成对生于叶腋，球形，直径1～1.2cm，无毛，梗纤弱。花果期3～12月。

Evergreen trees. Bark dark brown. Branchlets pubescent. Leaves leathery, oblanceolate, oblong to elliptic, 6-15(-20) cm × 2.5-6 cm, margin entire, midvein and lateralveins prominent abaxially. Figs yellow or reddish-brown, solitary or paired, axillary, globose, 1-1.2 cm in diam., glabrous, stalks slender. Fl. and fr. Mar.-Dec..

榕果 Multiple fruit
摄影：吴林芳 Photo by: Wu Linfang

叶背 Leaf abaxial surface
摄影：吴林芳 Photo by: Wu Linfang

叶 Leaf
摄影：吴林芳 Photo by: Wu Linfang

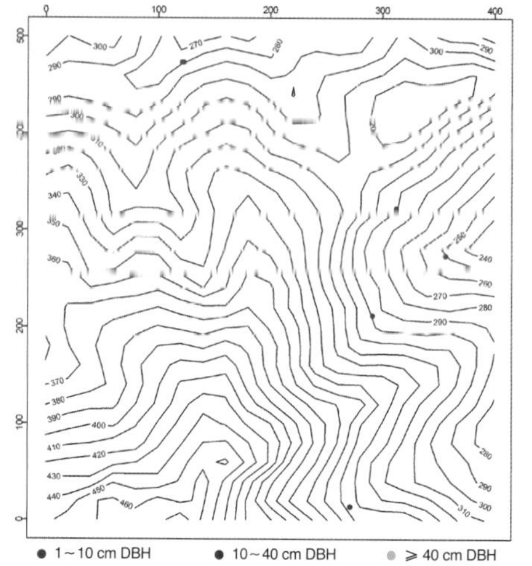

个体分布图 Distribution of individuals

径级分布表 DBH class

胸径等级 (Diameter class) (cm)	个体数 (No. of individuals in the plot)	比例 (Proportion) (%)
1～2	0	0.00
2～5	0	0.00
5～10	1	16.67
10～20	2	33.33
20～30	3	50.00
30～60	0	0.00
≥60	0	0.00

114 琴叶榕

qínyèróng | Fiddleleaf Fig

Ficus pandurata Hance
桑科 | Moraceae

代码（SpCode）= FICPAN
个体数（Individual number/20 hm^2）= 4
最大胸径（Max DBH）= 8.9 cm
重要值排序（Importance value rank）= 136

常绿灌木，高1～2m。叶纸质，提琴形或倒卵形，中间缢缩，叶面无毛，叶背脉上被疏毛或小突起，全缘。榕果单生于叶腋，鲜红色，椭圆形或球形，直径4～10mm，无毛或有粗糙的乳突，具纤弱果梗。花果期6～8月。

Evergreen shrubs, 1-2 m tall. Leaf blade obovate, or violin-shaped, with middle constricted, papery, abaxially sparsely pubescent on veins and with small cystoliths, adaxially glabrous, margin entire. Figs axillary on normal leafy shoots, solitary, scarlet, ellipsoid to globose, 4-10 mm in diam., glabrous or minutely scabrid-papillate; peduncles slender. Fl. and fr. Jun.-Aug..

榕果 Multiple fruit
摄影：吴林芳 photo by: Wu Linfang

叶 Leaf
摄影：吴林芳 Photo by: Wu Linfang

叶背 Leaf abaxial surface
摄影：吴林芳 Photo by: Wu Linfang

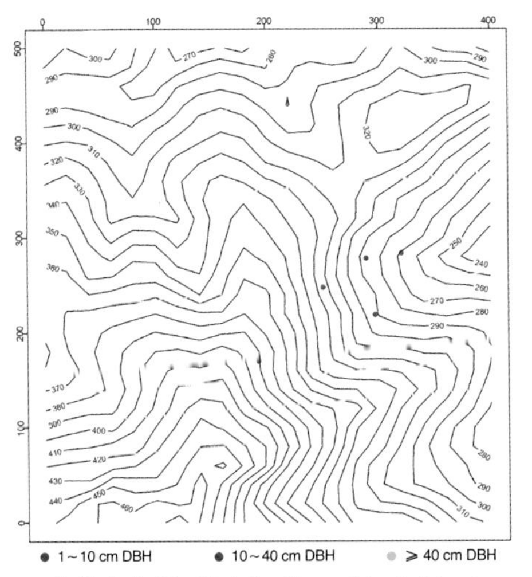

个体分布图 Distribution of individuals

径级分布表 DBH class

胸径等级 (Diameter class) (cm)	个体数 (No. of individuals in the plot)	比例 (Proportion) (%)
1～2	3	60.00
2～5	1	20.00
5～10	1	20.00
10～20	0	0.00
20～30	0	0.00
30～60	0	0.00
≥60	0	0.00

115 笔管榕

bǐguǎnróng | Japan Superb Fig

Ficus subpisocarpa Gagnep.

桑科 | Moraceae

代码（SpCode）= FICSUB

个体数（Individual number/20 hm^2）= 1

最大胸径（Max DBH）= 30.0 cm

重要值排序（Importance value rank）= 163

落叶乔木，有时有气生根。树皮黑褐色。小枝淡红色，无毛。托叶披针形约2cm；叶柄长3～7cm。叶近纸质，互生或簇生，椭圆形至长圆形，全缘或微波状。榕果紫黑色，单生或成对腋生或生于无叶的枝上，扁球形，直径5～8mm，果梗长3～4mm，花果期2～9月。

Deciduous trees, sometimes with aerial roots. Bark black-brown. Branchlets reddish, glabrous. Stipules lanceolate, ca. 2 cm. Petioles 3-7 cm. Leaves alternate or clustered, subpapery, elliptic to oblong, margin entire or slightly undurate. Figs purple-black, solitary or paired, axillary or on leafless branches, oblate, 5-8 mm in diam., peduncles 3-4 mm. Fl. and fr. Feb.-Sep..

叶背 Leaf abaxial surface

摄影：吴林芳 Photo by: Wu Linfang

枝叶 Branch and leaves

摄影：吴林芳 Photo by: Wu Linfang

果枝 Fruiting branch

摄影：吴林芳 Photo by: Wu Linfang

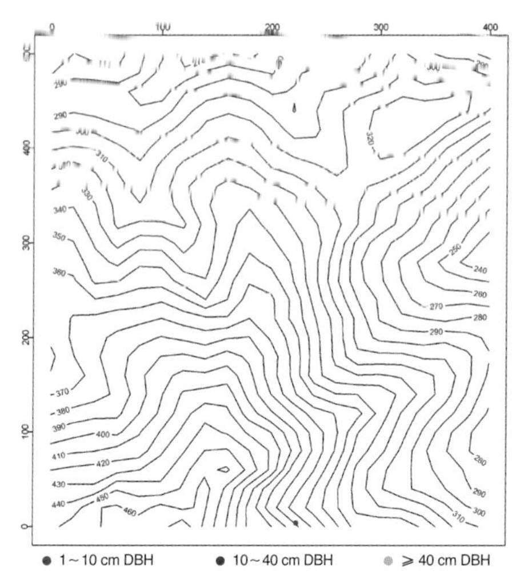

个体分布图 Distribution of individuals

径级分布表 DBH class

胸径等级 (Diameter class) (cm)	个体数 (No. of individuals in the plot)	比例 (Proportion) (%)
1～2	0	0.00
2～5	0	0.00
5～10	0	0.00
10～20	0	0.00
20～30	0	0.00
30～60	1	100.00
≥60	0	0.00

116 杂色榕（青果榕） zásèróng | Red-stem Fig

Ficus variegata Bl.
桑科 | Moraceae

代码（SpCode）= FIC1AR
个体数（Individual number/20 hm^2）= 13
最大胸径（Max DBH）= 19.9 cm
重要值排序（Importance value rank）= 108

常绿大乔木，高20m。树皮灰色。嫩枝绿色略被柔毛。叶柄5～7cm，叶厚纸质，阔卵形至卵状椭圆形，全缘，或略为波状及疏细齿，微被柔毛或无毛。榕果通常绿色，簇生于老茎发出的瘤状短枝上，梨形或球形，具梗，径1～2.5cm。花果期3～12月。

Evergreen large trees up to 20 m tall. Bark grey. Young branchlets green, sparse pubescent. Petioles 5-7 cm. Leaves thickly papery, broadly ovate to ovate-elliptic, margin entire, shallowly sinuate or sparsely serrulate, sparsely pubescent or glabrous. Figs usually green, clustered on brachyblasts of trunk or branch, pear-shaped or globose, 1-2.5 cm in diam.. Fl. and fr. Mar.-Dec..

果枝 Fruiting branch
摄影：吴林芳 Photo by: Wu Linfang

叶 Leaf
摄影：吴林芳 Photo by: Wu Linfang

叶背 Leaf abaxial surface
摄影：吴林芳 Photo by: Wu Linfang

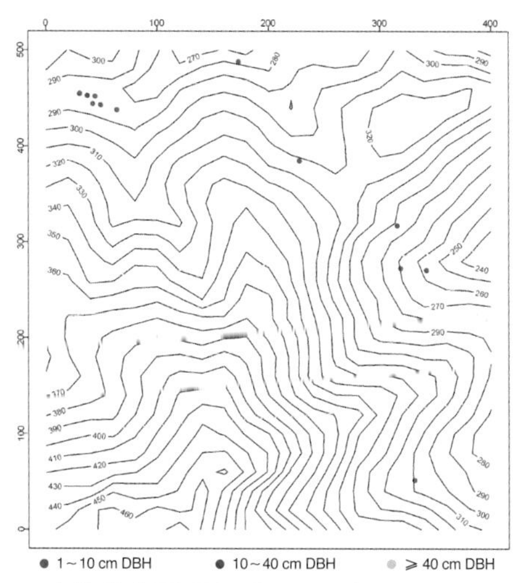

● 1～10 cm DBH ● 10～40 cm DBH ● ≥40 cm DBH
个体分布图 Distribution of individuals

径级分布表 DBH class

胸径等级 (Diameter class) (cm)	个体数 (No. of individuals in the plot)	比例 (Proportion) (%)
1～2	4	30.77
2～5	5	38.46
5～10	2	15.38
10～20	2	15.38
20～30	0	0.00
30～60	0	0.00
≥60	0	0.00

117 变叶榕

biànyèróng | Variedleaf Fig

Ficus variolosa Lindl. ex Benth.
桑科 | Moraceae

代码（SpCode）= FIC2AR
个体数（Individual number/20 hm^2）= 74
最大胸径（Max DBH）= 12.5 cm
重要值排序（Importance value rank）= 63

常绿或半常绿灌木或小乔木，高3～10m。树皮灰褐色。小枝节间短。叶薄革质，狭椭圆形、椭圆状披针形，全缘。榕果成对或单生于叶腋，球形，径1～1.2cm，表面有瘤体，果梗长8～12mm。花果期2～12月。

Evergreen or semi-evergreen shrubs or small trees, 3-10 m tall. Bark grey-brown. Branchlets with short internodes. Leaves thinly leathery, narrowly ellipitic, ellipitic-lanceolate, margin entire. Figs paired or solitary, axillary, globose, 10-12 mm in diam., verrucose, peduncles 8-12 mm. Fl. and fr. Feb.-Dec..

野外识别特征：

1. 树皮灰褐色，杆上不着生榕果；
2. 叶薄革质，多种形态，全缘；
3. 榕果成对或单个生于叶腋，球形，黑色。

Key notes for identification:

1. Bark grey-brown, no figs on trunk.
2. Leaves thinly leathery multiform, margin entire.
3. Figs axillary or solitary, globose, black.

果枝 Branch and leaves
摄影：吴林芳 Photo by: Wu Linfang

果 Fruit
摄影：吴林芳 Photo by: Wu Linfang

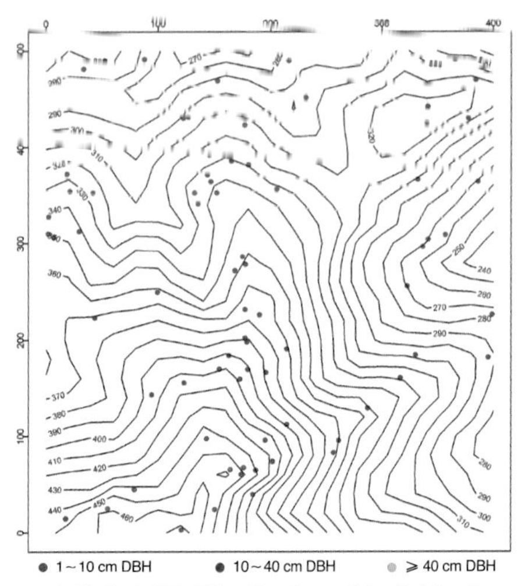

个体分布图 Distribution of individuals

径级分布表 DBH class

胸径等级 (Diameter class) (cm)	个体数 (No. of individuals in the plot)	比例 (Proportion) (%)
1～2	18	24.32
2～5	36	48.65
5～10	17	22.97
10～20	3	4.05
20～30	0	0.00
30～60	0	0.00
≥60	0	0.00

118 白肉榕

báiròuróng | White Fig

Ficus vasculosa Wall. ex Miq.
桑科 | Moraceae

代码（SpCode）= FICVAS
个体数（Individual number/20 hm^2）= 7
最大胸径（Max DBH）= 11.8 cm
重要值排序（Importance value rank）= 117

常绿乔木，高10～15m。树皮灰色平滑。小枝灰褐色。叶革质，椭圆形至长披针形，长4～11cm，宽2～4cm，全缘或不规则分裂，叶脉在两面明显。榕果熟时黄色或红黄色，成对生于叶腋，球形，径7～10 mm，果梗7～8 mm。花果期2～12月。

Evergreen trees, 10-15 m tall. Bark grey, smooth. Branchlets grey-brown. Leaves leathery, elliptic to oblong-lanceolate, 4-11 cm × 2-4 cm, margin entire or irregularly lobed, veins prominent on both surfaces. Figs yellow or yellow-red when mature, paired, axillary, globose, 7-10 mm in diam., peduncles 7-8 mm. Fr. and fr. Feb.-Dec..

野外识别特征：

1. 树皮灰色，光滑，不着生榕果；
2. 叶革质，椭圆形至长披针形，先端钝或渐尖；
3. 榕果熟时黄色或红黄色，球形。

Key notes for identification:

1. Bark grey, smooth, no figs on trunk.
2. Leaves leathery, elliptic to oblong-lanceolate, apex obtuse to acuminate.
3. Figs yellow or yellow-red when mature, globose.

果枝 Fruiting branch
摄影：吴林芳 Photo by: Wu Linfang

叶 Leaf
摄影：吴林芳 Photo by: Wu Linfang

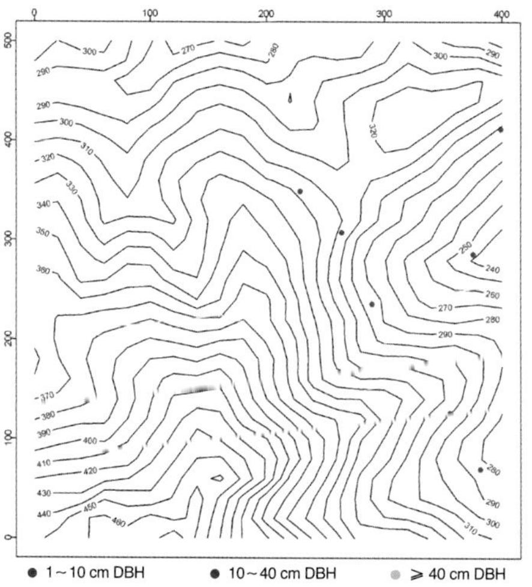

个体分布图 Distribution of individuals

径级分布表 DBH class

胸径等级 (Diameter class) (cm)	个体数 (No. of individuals in the plot)	比例 (Proportion) (%)
1～2	0	0.00
2～5	2	28.57
5～10	2	28.57
10～20	3	42.86
20～30	0	0.00
30～60	0	0.00
≥60	0	0.00

119 紫麻

zǐmá | Woodnettle

Oreocnide frutescens (Thunb.) Miq.
荨麻科 | Urticaceae

代码（SpCode）= OREFRU
个体数（Individual number/20 hm^2）= 1
最大胸径（Max DBH）= 2.1 cm
重要值排序（Importance value rank）= 182

落叶灌木或小乔木。小枝和叶柄紫褐色。叶常生枝上部，草质或纸质，叶长3～15cm，宽1.5～6cm，基出脉3，边缘或下部以上具齿，顶部齿较尖，叶疏被硬毛。花序几无梗或具分叉短梗，团伞状花簇生，雌雄异株。瘦果卵球形，长约1.2mm。花期2～4月，果期6～10月。

Deciduous shrubs or small trees. Branchlets and petioles purplish brown. Leaves always on top of branch, leaf blade 3-15 cm × 1.5-6 cm, herbaceous or papery, 3-veined, margin serrate or dentate from base or sometimes middle, apex acuminate. Sparsely strigillose. Inflorescences almost sessile clusters or pedunculate dichotomously branched cymes. Achene ovoid, ca. 1.2 mm. Fl. Feb.-Apr., fr. Jun.-Oct..

叶 Leaf
摄影：吴林芳 Photo by: Wu Linfang

果 Fruit
摄影：吴林芳 Photo by: Wu Linfang

花枝 Flowering branch
摄影：吴林芳 Photo by: Wu Linfang

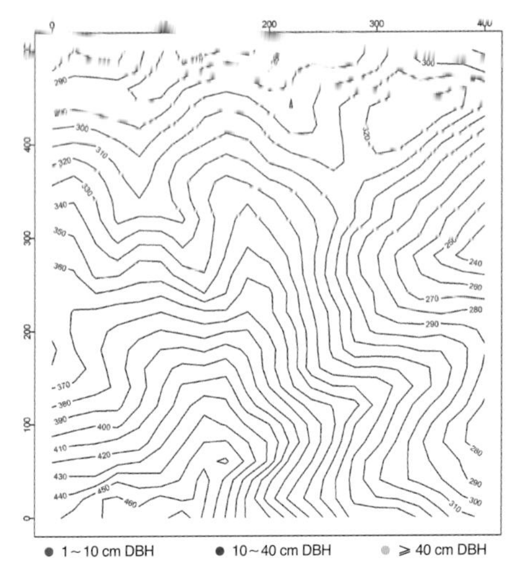

个体分布图 Distribution of individuals

径级分布表 DBH class

胸径等级 (Diameter class) (cm)	个体数 (No. of individuals in the plot)	比例 (Proportion) (%)
1～2	0	0.00
2～5	1	100.0
5～10	0	0.00
10～20	0	0.00
20～30	0	0.00
30～60	0	0.00
≥60	0	0.00

120 凹叶冬青

aoyèdōngqīng | Emarginate Holly

Ilex championii Loes.

冬青科 | Aquifoliaceae

代码（SpCode）= ILECHA

个体数（Individual number/20 hm^2）= 41

最大胸径（Max DBH）= 20 cm

重要值排序（Importance value rank）= 82

常绿灌木或乔木，高13（~15）m。树皮灰白色或灰褐色。当年生幼枝具纵棱槽，被微毛，紫褐色。叶厚革质，卵形或倒卵形，先端圆而微凹或微缺或短突尖，两面无毛，全缘，叶背具深色腺点。聚伞花序簇生于叶腋。果扁球形，熟时红色，直径3~4mm。分核4，椭圆状倒卵形。花期6~7月，果期8~11月。

Evergreen shrubs or trees, 13 (-15) m tall. Bark gray-white or gray-brown. Current year branchlets purple-brown, longitudinally ridged and puberulent. Leaf blade ovate or obovate, thickly leathery, apex retuse or emarginate, or shortly and abruptly acuminate, both surfaces glabrous, margin entire, abaxially deeply colored punctate. Cymes, fasciculate, axillary on first to second year's branchlets. Fruit red when mature, compressed globose, 3-4 mm in diam.. Pyrenes 4, ellipsoidal-obovoid. Fl. Jun.-Jul., fr. Aug.-Nov..

野外识别特征：

1. 叶厚革质，先端圆而微凹或微缺或短突尖，两面无毛；
2. 叶全缘，叶背具深色腺点；
3. 当年生幼枝具纵棱槽，被微毛，紫褐色。

Key notes for identification:

1. Leaf blade thickly leathery, apex retuse or emarginate, or shortly and abruptly acuminate.
2. Leaf blade margin entire, abaxially deeply colored punctate.
3. Current year branchlets purple-brown, sulcate, longitudinally ridged and sulcate, puberulent.

叶 Leaf

摄影：吴林芳 Photo by: Wu Linfang

叶背 Leaf abaxial surface

摄影：吴林芳 Photo by: Wu Linfang

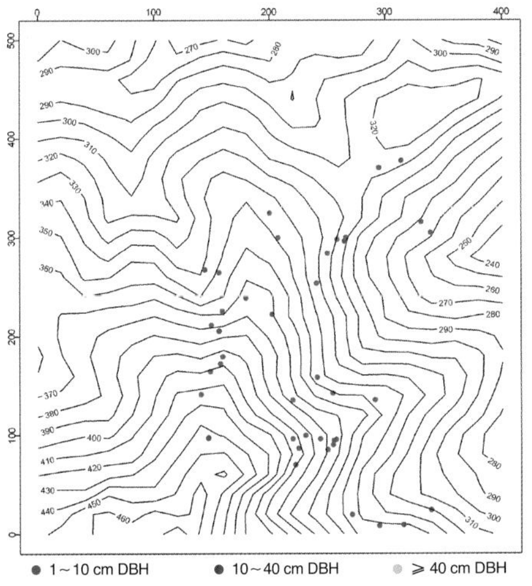

个体分布图 Distribution of individuals

径级分布表 DBH class

胸径等级 (Diameter class) (cm)	个体数 (No. of individuals in the plot)	比例 (Proportion) (%)
1~2	9	21.95
2~5	15	36.59
5~10	8	19.51
10~20	8	19.51
20~30	1	2.44
30~60	0	0.00
≥60	0	0.00

121 沙坝冬青 shābàdōngqīng | Chapae Holly

Ilex chapaensis Merr.
冬青科 | Aquifoliaceae

代码（SpCode）= ILECHA
个体数（Individual number/20 hm^2）= 126
最大胸径（Max DBH）= 61.0 cm
重要值排序（Importance value rank）= 42

落叶乔木，高达20m。小枝幼时被毛后变无毛，皮孔明显，短枝少数。叶纸质，卵状椭圆形或长椭圆形，边缘具浅圆齿，无毛。花雌雄异株，6～8基数。果熟时绿色，径1.2～2cm。分核6或7，宿存柱头圆柱状。花期4月，果期10～11月。

Deciduous trees, to 20 m tall. Branchlets glabrescent. Lenticels conspicuous. Abbreviated shoots few. Leaves papery. Leaf blade ovate-elliptic or oblong-elliptic, margin crenulate-serrulate, glabrous. Flowers 6-8-merous. Unisexual. Fruit green when mature, 1.2-2 cm in diam.. Persistent stigma columnar. Pyrenes 6 or 7. Fl. Apr., fr. Oct.-Nov..

野外识别特征：

1. 落叶乔木，树皮灰白色；
2. 叶纸质，边缘具浅圆齿，无毛；先端短渐尖或钝，少数圆形；
3. 果熟时绿色，宿存柱头圆柱状，分核6或7。

Key notes for identification:

1. Deciduous trees, bark gray.
2. Leaves papery, margin crenulate-serrulate, glabrous; apex shortly acuminate or obtuse, rarely rounded.
3. Fruit green when mature persistent stigma columnar, pyrenes 6 or 7.

叶 Leaf
摄影：吴林芳 Photo by: Wu Linfang

叶背 Leaf abaxial surface
摄影：吴林芳 Photo by: Wu Linfang

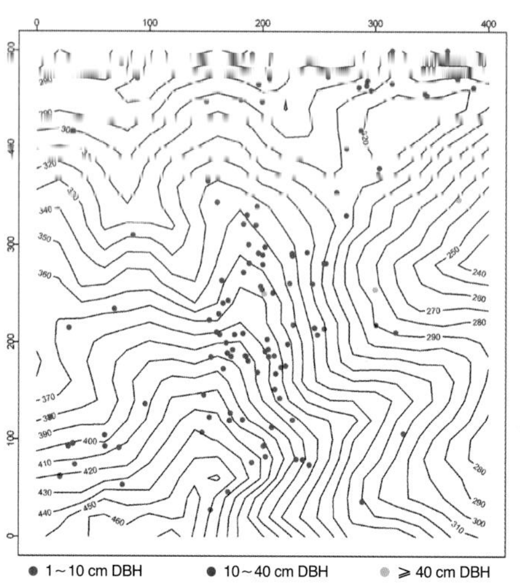

个体分布图 Distribution of individuals

径级分布表 DBH class

胸径等级 (Diameter class) (cm)	个体数 (No. of individuals in the plot)	比例 (Proportion) (%)
1～2	2	1.59
2～5	9	7.14
5～10	20	15.87
10～20	45	35.71
20～30	42	33.33
30～60	7	5.56
≥60	1	0.79

126 铁冬青 tiědōngqīng I Chinese Holly

Ilex rotunda Thunb.
冬青科 I Aquifoliaceae

代码（SpCode）= ILEROT
个体数（Individual number/20 hm^2）= 15
最大胸径（Max DBH）= 27.2 cm
重要值排序（Importance value rank）= 94

常绿乔木，高达20m。树皮灰色至灰黑色。幼枝具纵棱，无毛，皮孔不明显。叶薄革质或纸质，卵形、长圆形或椭圆形，两面无毛，全缘。聚伞花序单生于当年生叶腋。果红色，近球形，直径4～6mm；分核5～7，椭圆形，具3条纹2沟槽。花期4～6月，果期8～12月。

Evergreen trees up to 20 m tall. Bark gray to gray-black. Young branchlets longitudinally angular, glabrous, lenticels inconspicuous. Leaf blade ovate, obovate, or elliptic, thinly leathery or papery, both surfaces glabrous, margin entire. Cymes solitary axillary on first year's branchlets. Fruit red, subglobose, 4-6 mm in diam.. Pyrenes 5-7, ellipsoidal, 3-striate and 2-sulcate. Fl. Apr.-Jun., fr. Aug.-Dec..

果 Fruits
摄影：吴林芳 Photo by: Wu Linfang

花枝 Flowering branch
摄影：吴林芳 Photo by: Wu Linfang

果枝 Fruiting branch
摄影：吴林芳 Photo by: Wu Linfang

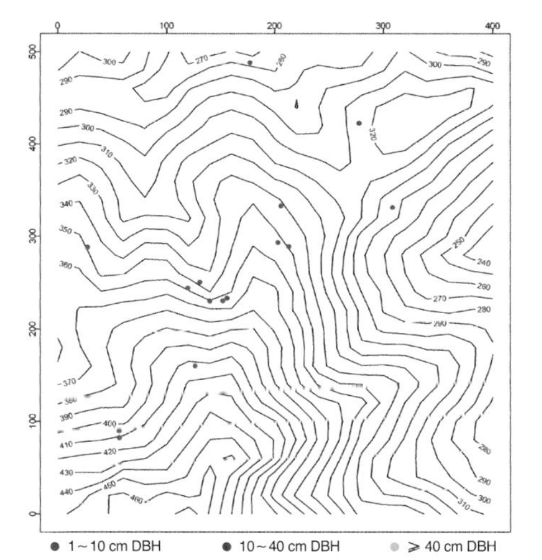

个体分布图 Distribution of individuals

径级分布表 DBH class

胸径等级 (Diameter class) (cm)	个体数 (No. of individuals in the plot)	比例 (Proportion) (%)
1～2	0	0.00
2～5	0	0.00
5～10	2	13.33
10～20	7	46.67
20～30	6	40.00
30～60	0	0.00
≥60	0	0.00

127 三花冬青

sānhuādōngqīng | Tri-flowered Holly

Ilex triflora Bl.

冬青科 | Aquifoliaceae

代码（SpCode）= ILETRI

个体数（Individual number/20 hm²）= 183

最大胸径（Max DBH）= 58.0 cm

重要值排序（Importance value rank）= 51

常绿灌木或乔木，高2～10m。幼枝“之”字形，近四棱，幼时被毛；无皮孔。叶近革质，椭圆形、长圆形、卵状椭圆形或倒卵形，边缘细圆齿。花淡紫色，4基数，簇生于2年生枝叶腋。果黑色，球形，径4～7mm。分核4，卵状椭圆形，具3条纹，无沟槽。花期5～6月，果期8～11月。

Evergreen shrubs or trees, 2-10 m tall. Branchlets zigzag, subquadrangular, pubescent when young, lenticels absent. Leaf blade elliptic, oblong, ovate-elliptic, obovate, subleathery, margin crenate-serrate. Flowers lavender, 4-merous, fasciculate in leaf axils of second year’s branchlets. Fruit black, globose, 4-7 mm in diam.. Pyrenes 4, ovoid-ellipsoidal, 3-striate, not sulcate. Fl. May-Jul., fr. Aug.-Nov..

果枝 Fruiting branch

摄影：吴林芳 Photo by: Wu Linfang

叶背 Leaf abaxial surface

摄影：吴林芳 Photo by: Wu Linfang

叶 Leaf

摄影：吴林芳 Photo by: Wu Linfang

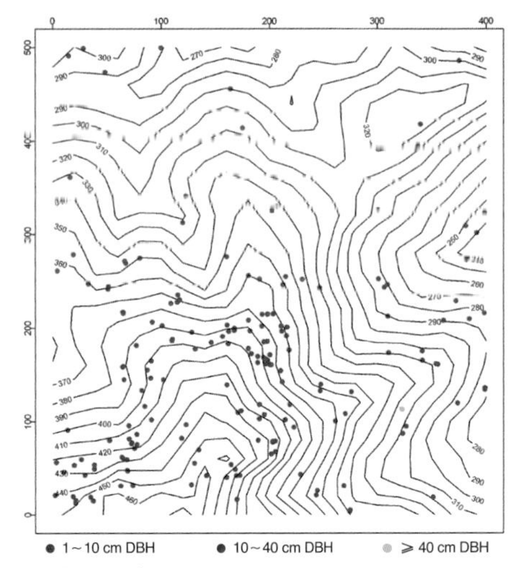

个体分布图 Distribution of individuals

径级分布表 DBH class

胸径等级 (Diameter class) (cm)	个体数 (No. of individuals in the plot)	比例 (Proportion) (%)
1～2	48	26.23
2～5	75	40.98
5～10	36	19.67
10～20	18	9.84
20～30	3	1.64
30～60	3	1.64
≥60	0	0.00

128 疏花卫矛 shuhuāwèimáo | Loose-flowered Euonymus

Euonymus laxiflorus Champ. ex Benth.
卫矛科 | Celastraceae

代码（SpCode）= EUOLAX
个体数（Individual number/20 hm²）= 304
最大胸径（Max DBH）= 6.8 cm
重要值排序（Importance value rank）= 40

落叶灌木或小乔木，高3～12m。小枝灰绿色，圆柱形，嫩枝绿色。单叶对生，叶柄短，3～5mm；叶薄革质，倒卵状椭圆形或卵形，近全缘。聚伞花序分枝疏松，5～9朵花，花5基数。蒴果倒卵形，5棱沟，鲜时桃红色。种子卵形，棕褐色，被橙色的假种皮部分覆盖。花期3～8月，果期5～11月。

Deciduous shrubs to small trees, 3-12 m tall. Branches greenish gray, terete; twigs greenish. Leaves opposite. Petioles 3-5 mm. Leaf blade thinly leathery, elliptic-obovate or ovate, margin nearly entire. Cymes loose, 5-9-flowered, 5- merous. Capsule obovoid, 5-angled and grooved, pinkish or reddish when fresh. Seeds ovoid, dark brown, partially covered by orange aril. Fl. Mar.-Aug., fr. May-Nov..

果 Fruit
摄影：吴林芳 Photo by: Wu Linfang

花序 Inflorescence
摄影：吴林芳 Photo by: Wu Linfang

果枝 Fruiting branch
摄影：吴林芳 Photo by: Wu Linfang

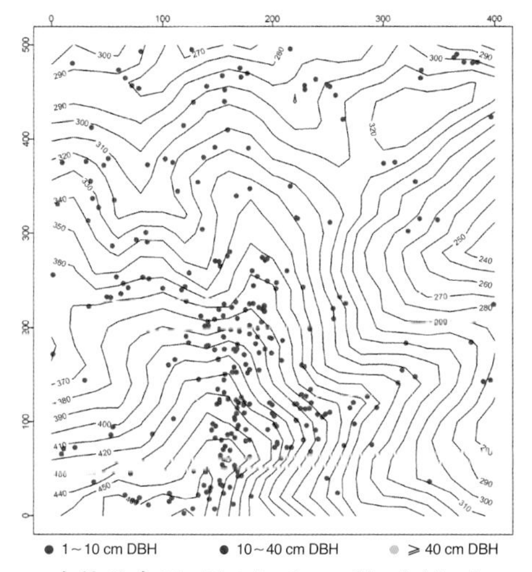

个体分布图 Distribution of individuals

径级分布表 DBH class

胸径等级 (Diameter class) (cm)	个体数 (No. of individuals in the plot)	比例 (Proportion) (%)
1～2	234	76.97
2～5	68	22.37
5～10	2	0.66
10～20	0	0.00
20～30	0	0.00
30～60	0	0.00
≥60	0	0.00

129 中华卫矛

zhōnghuáwèimáo | Chinese Euonymus

Euonymus nitidus Benth.
卫矛科 | Celastraceae

代码（SpCode）= EUONIT
个体数（Individual number/20 hm²）= 1
最大胸径（Max DBH）= 1.2 cm
重要值排序（Importance value rank）= 191

常绿灌木或小乔木，高2～10m。小枝灰黑至灰褐色，圆柱形，粗壮，嫩枝绿色或黄绿色，具细槽。叶柄长，5～8（～12）mm；叶革质或厚纸质，椭圆形或长椭圆形。聚伞花序比叶短，多花，4基数。蒴果三角卵形，桃红色。种子卵形，棕褐色，假种皮橙色。花期3～7月，果期7月至翌年1月。

Evergreen shrubs or small trees, 2-10 m tall. Branches gray-black to gray-brown, terete, sturdy, twigs greenish or yellow-greenish, striate. Petiole 5-8 (-12) mm, leaf blade leathery or thickly papery, elliptic or oblong-elliptic. Cymes shorter than leaves, 1-15-flowered; flowers 4-merous. Capsule deltoid-ovoid, peachblow when fresh. Seeds ovoid, dark brown; aril orange. Fl. Mar.-Jul., fr. Jul.-Jan. of next year.

果 Fruit
摄影：吴林芳 Photo by: Wu Linfang

果枝 Fruiting branch
摄影：吴林芳 Photo by: Wu Linfang

叶 Leaf
摄影：吴林芳 Photo by: Wu Linfang

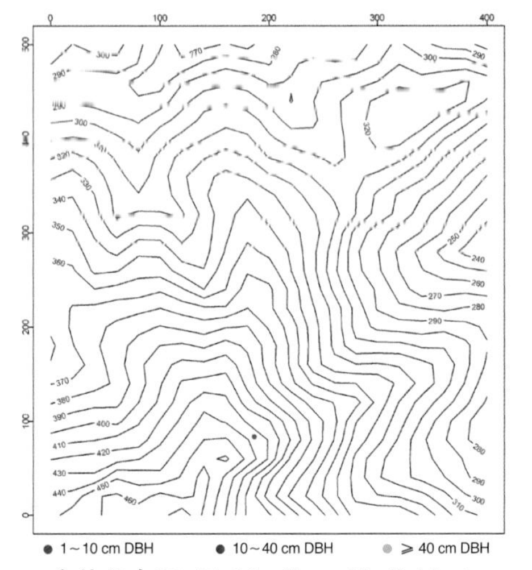

个体分布图 Distribution of individuals

径级分布表 DBH class

胸径等级 (Diameter class) (cm)	个体数 (No. of individuals in the plot)	比例 (Proportion) (%)
1～2	1	100.00
2～5	0	0.00
5～10	0	0.00
10～20	0	0.00
20～30	0	0.00
30～60	0	0.00
≥60	0	0.00

134 柑橘 gānjú | Glycosmis

Citrus reticulata Blanco
芸香科 | Rutaceae

代码（SpCode）= CITRET
个体数（Individual number/20 hm^2）= 5
最大胸径（Max DBH）= 9.5 cm
重要值排序（Importance value rank）= 140

常绿小乔木。树皮灰褐色，多刺，枝具棱，少刺。叶通常为单身复叶，叶缘上部具钝齿，极少全缘，先端微凹。花单朵至3朵簇生叶腋或枝顶。果淡黄色至橙色。花期4～5月，果期10～12月。

Evergreen small trees. Bark gary-brown, more spines, branchlets with few spines, angular. Leaves usually unifoliate compound, margin apically obtusely crenulate or rarely entire, apex emarginate. Flowers solitary to 3 in a fascicle. Fruit pale yellow to orange. Fl. Apr.-May, fr. Oct.-Dec..

树干 Trunk
摄影：吴林芳 Photo by: Wu Linfang

柑果 Hesperidium
摄影：卓书斌 Photo by: Zuo Shubig

叶 Leaf
摄影：吴林芳 Photo by: Wu Linfang

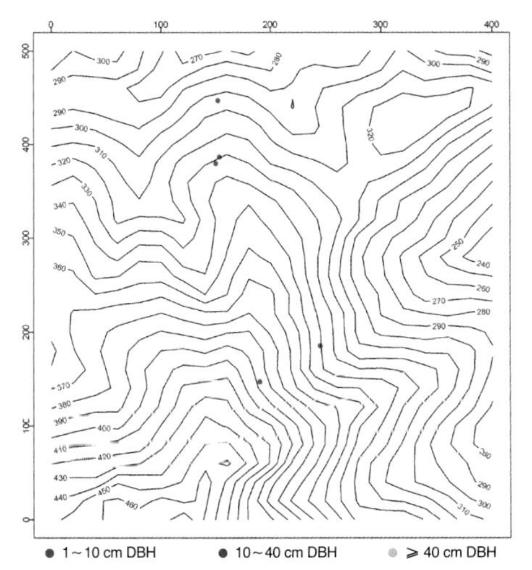

个体分布图 Distribution of individuals

径级分布表 DBH class

胸径等级 (Diameter class) (cm)	个体数 (No. of individuals in the plot)	比例 (Proportion) (%)
1～2	1	20.00
2～5	3	60.00
5～10	1	20.00
10～20	0	0.00
20～30	0	0.00
30～60	0	0.00
≥60	0	0.00

135 小花山小橘 xiǎohuāshānxiǎojú | Glycosmis

Glycosmis parviflora (Sims) Little
芸香科 | Rutaceae

代码（SpCode）= GLYPAR
个体数（Individual number/20 hm^2）= 8
最大胸径（Max DBH）= 9.5 cm
重要值排序（Importance value rank）= 141

常绿灌木，高1～3m。2～4小叶，有时单叶或具5小叶；叶互生，小叶椭圆形、长圆形或披针形，无毛，全缘。聚伞圆锥花序顶生或腋生。浆果黄白色，后变深红色，球形或椭圆形。花期3～5月，果期7～9月。

Evergreen shrubs, 1-3 m tall. Leaves (1 or) 2-4 (or 5)-foliolate, alternate, leaflet blades elliptic, oblong, or lanceolate, glabrous, margin entire. Thyrse axillary or terminal, Berry pale yellowish white but turning reddish to dark vermilion, globose to ellipsoid. Fl. Mar.-May, fr. Jul.-Sep..

花枝 Flowering branch
摄影：吴林芳 Photo by: Wu Linfang

果枝 Fruiting branch
摄影：吴林芳 Photo by: Wu Linfang

果 Fruit
摄影：吴林芳 Photo by: Wu Linfang

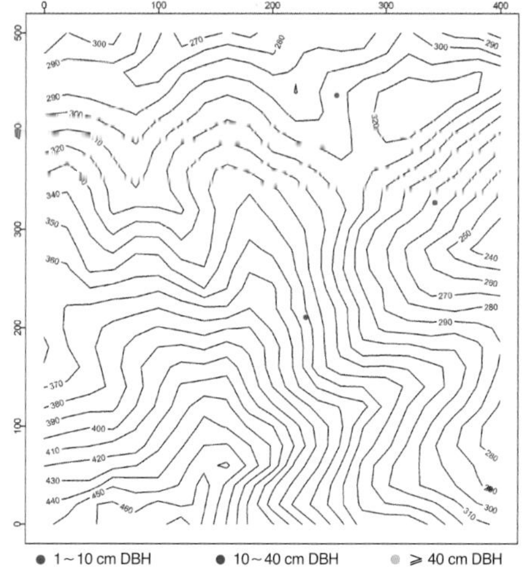

个体分布图 Distribution of individuals

径级分布表 DBH class

胸径等级 (Diameter class) (cm)	个体数 (No. of individuals in the plot)	比例 (Proportion) (%)
1～2	2	25.00
2～5	5	62.50
5～10	1	12.50
10～20	0	0.00
20～30	0	0.00
30～60	0	0.00
≥60	0	0.00

136 三椏苦 sānyākǔ | Thin Evodia

Melicope pteleifolia (Champ. ex Benth.) T. G. Hartley
芸香科 | Rutaceae

代码（SpCode）= MELPTE
个体数（Individual number/20 hm^2）= 62
最大胸径（Max DBH）= 5.1 cm
重要值排序（Importance value rank）= 71

半常绿灌木或乔木，高1～14m。雌雄异株，少数雌雄同花。叶具3小叶，少数单叶，对生，小叶卵状椭圆形、椭圆形或椭圆状倒卵形，全缘具波浪。聚伞花序腋生，少数顶生。果具1～4个小囊，小囊果近球形至倒卵形。花期4～5月，果期8～9月。

Semievergreen shrubs or trees, 1-14 m tall. Dioecious or rarely monoclinous. Leaves 3-foliolate (occasional leaves 1-foliolate), opposite, leaflet blades ovate-elliptic, elliptic, elliptic-obovate, margin entire and mavy. Thyrse axillary, rarely terminal. Fruit 1-4 basally connate follicles, follicles subglobose to obovoid. Fl. Apr.-May, fr. Aug.-Sep..

花枝 Flowering branch
摄影：吴林芳 Photo by: Wu Linfang

叶 Leaf
摄影：吴林芳 Photo by: Wu Linfang

果枝 Fruiting branch
摄影：吴林芳 Photo by: Wu Linfang

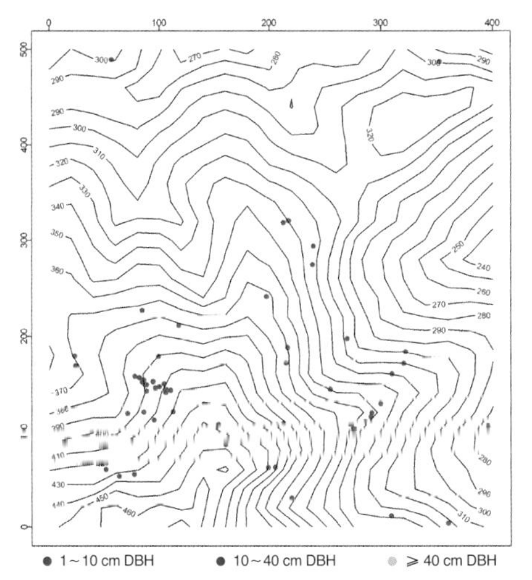

个体分布图 Distribution of individuals

径级分布表 DBH class

胸径等级 (Diameter class) (cm)	个体数 (No. of individuals in the plot)	比例 (Proportion) (%)
1～2	41	66.13
2～5	20	32.26
5～10	1	1.61
10～20	0	0.00
20～30	0	0.00
30～60	0	0.00
≥60	0	0.00

137 簕棁花椒

lèdǎnghuājiāo | Prickly Ash

Zanthoxylum avicennae (Lam.) DC
芸香科 | Rutaceae

代码（SpCode）= ZANAVI
个体数（Individual number/20 hm^2）= 1
最大胸径（Max DBH）= 16.1 cm
重要值排序（Importance value rank）= 171

落叶乔木，高15m。树皮具刺。小枝和叶无毛。小叶11～21片，叶轴具翼；小叶片对生或不规则对生，歪卵形、长菱形、倒卵形或镰形，具油腺点。圆锥花序顶生，多花；花轴带紫的浅红色。花5基数。小囊果带紫的浅红色，多被腺点。花期6～8月，果期10～12月。

Deciduous trees, to 15 m tall. Bark with prickle. Branchlets and leaves glabrous, with prickles. Leaves 11-21-foliolate, rachis winged, leaflet blades opposite or subopposite, obliquely ovate, rhomboidal, obovate, or falcate, with oil glands. Panicle terminal, many flowered, rachis purplish red. Flowers 5-merous. Follicles pale purplish red, with many glandular dots. Fl. Jun.-Aug., fr. Oct.-Dec..

树干 Trunk
摄影：吴林芳 Photo by: Wu Linfang

果枝 Fruiting branch
摄影：吴林芳 Photo by: Wu Linfang

花枝 Flowering branch
摄影：吴林芳 Photo by: Wu Linfang

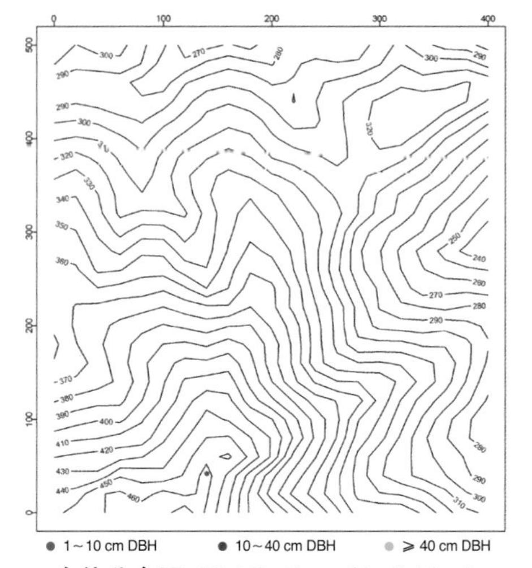

个体分布图 Distribution of individuals

径级分布表 DBH class

胸径等级 (Diameter class) (cm)	个体数 (No. of individuals in the plot)	比例 (Proportion) (%)
1～2	0	0.00
2～5	0	0.00
5～10	0	0.00
10～20	1	100.00
20～30	0	0.00
30～60	0	0.00
≥60	0	0.00

138 大叶臭花椒 dàyèchòuhuājiāo | Bigleaf Pricklyash

Zanthoxylum myriacanthum Wall. Ex Hook. f.
芸香科 | Rutaceae

代码（SpCode）= ZANMYR
个体数（Individual number/20 hm^2）= 17
最大胸径（Max DBH）= 18.8 cm
重要值排序（Importance value rank）= 97

落叶乔木，高达15m。枝聚生树顶，基部树皮及花序轴具刺。叶无刺，7～17小叶，小叶对生，阔卵形、卵状椭圆形或长圆形，两面无毛，油腺点较大。圆锥花序顶生，多花，花5基数。小果囊棕红色，具油腺点。花期6～8月，果期9～11月。

Deciduous trees, to 15 m tall. Branchlets toward apex, base bark and rachis of inflorescences with prickles. Leaves without prickles, 7-17-foliolate, leaflet blades opposite, broadly ovate, ovate-elliptic, or oblong, both surfaces glabrous, oil glands numerous, large. Panicle terminal, many flowered. Flowers 5-merous. Fruit follicles reddish brown, oil glands numerous. Fl. Jun.-Aug, fr. Sep.-Nov..

树干 Trunk
摄影：吴林芳 Photo by: Wu Linfang

花枝 Flowering branch
摄影：吴林芳 Photo by: Wu Linfang

叶 Leaf
摄影：吴林芳 Photo by: Wu Linfang

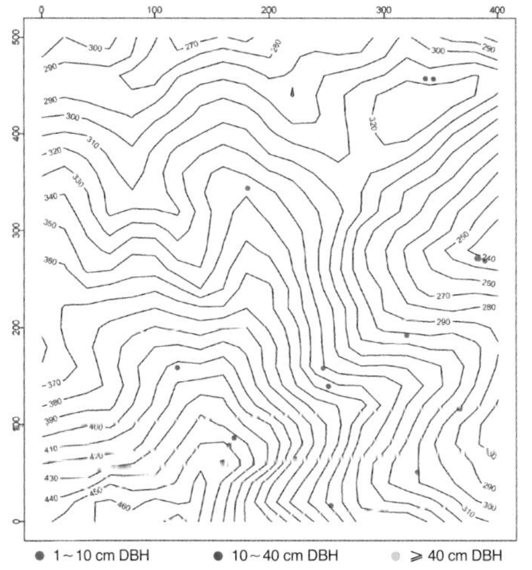

个体分布图 Distribution of individuals

径级分布表 DBH class

胸径等级 (Diameter class) (cm)	个体数 (No. of individuals in the plot)	比例 (Proportion) (%)
1～2	1	5.88
2～5	7	41.18
5～10	3	17.65
10～20	6	35.29
20～30	0	0.00
30～60	0	0.00
≥60	0	0.00

139 橄榄

gǎnlǎn | Chinese White Olive

Canarium album (Loureiro) Raeuch.
橄榄科 | Burseraceae

代码（SpCode）= CANALB
个体数（Individual number/20 hm^2）= 411
最大胸径（Max DBH）= 50.7 cm
重要值排序（Importance value rank）= 36

常绿大乔木，高7～25（～35）m。叶互生，奇数羽状复叶，具托叶，早落；小叶3～6对，小叶披针形、椭圆形或卵形，全缘。花序腋生。核果卵形或纺锤形，长25～35mm，黄绿色，无毛。花期4～5月，果期10～12月。

Great evergreen trees, 7-25 (-35) m tall. Leaves alternate, odd-pinnate, with stipulate; leaflets 3-6 pairs, blades lanceolate, elliptic, or ovate, margin entire. Inflorescences axillary. Drupe ovoid or spindle-shaped, 25-35 mm, yellow-green, glabrous. Fl. Apr.-May, fr. Oct.-Dec..

花枝 Flowering branch
摄影：吴林芳 Photo by: Wu Linfang

果枝 Fruiting branch
摄影：吴林芳 Photo by: Wu Linfang

叶 Leaf
摄影：吴林芳 Photo by: Wu Linfang

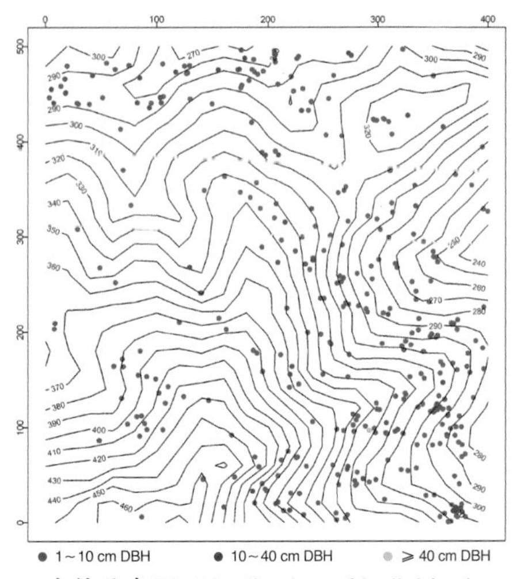

个体分布图 Distribution of individuals

径级分布表 DBH class

胸径等级 (Diameter class) (cm)	个体数 (No. of individuals in the plot)	比例 (Proportion) (%)
1～2	147	39.20
2～5	126	33.60
5～10	48	12.80
10～20	34	9.07
20～30	16	4.27
30～60	4	1.07
≥60	0	0.00

140 乌榄

wūlǎn | Chinese Black Olive

Canarium pimela K. D. Koenig
橄榄科 | Burseraceae

代码（SpCode）= CANTRA
个体数（Individual number/20 hm^2）= 6
最大胸径（Max DBH）= 6.9 cm
重要值排序（Importance value rank）= 127

常绿乔木，高20m。叶对生，奇数羽状复叶，无托叶；小叶4～6对，小叶阔椭圆形、卵形或成圆形，无毛，全缘。核果熟时紫黑色，狭卵形，长3～4cm，宽1.7～2cm，横切面近圆形。花期4～5月，果期5～11月。

Evergreen trees, to 20 m tall. Leaves opposite, odd-pinnate, exstipulate, leaflets 4-6 pairs, blades broadly elliptic, ovate, or rounded, glabrous, margin entire. Drupe purple-black when ripe, narrowly ovoid, 3-4 cm × 1.7-2 cm, cross section nearly rounded. Fl. Apr.-May, fr. May-Nov..

叶背 Leaf abaxial surface
摄影：吴林芳 Photo by: Wu Linfang

果枝 Fruiting branch
摄影：吴林芳 Photo by: Wu Linfang

果 Fruit
摄影：吴林芳 Photo by: Wu Linfang

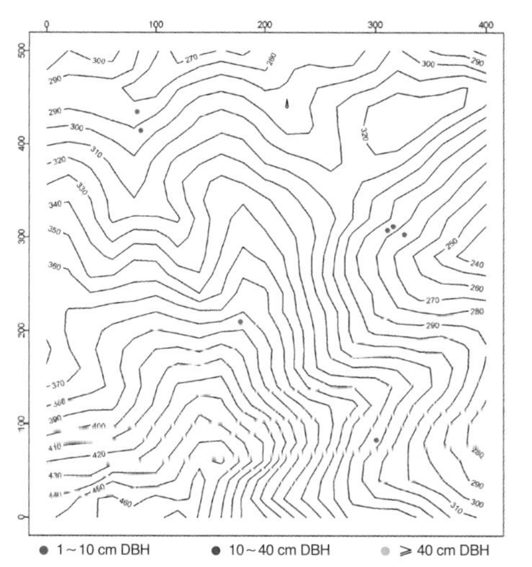

个体分布图 Distribution of individuals

径级分布表 DBH class

胸径等级 (Diameter class) (cm)	个体数 (No. of individuals in the plot)	比例 (Proportion) (%)
1～2	2	33.33
2～5	3	50.00
5～10	1	16.67
10～20	0	0.00
20～30	0	0.00
30～60	0	0.00
≥60	0	0.00

141 龙眼 lóngyǎn | Longan

Dimocarpus longan Lour.
无患子科 | Sapindaceae

代码（SpCode）= DIMLON
个体数（Individual number/20 hm^2）= 1
最大胸径（Max DBH）= 2.0 cm
重要值排序（Importance value rank）= 193

常绿乔木。树皮粗糙，纵裂。偶数羽状复叶互生，小叶对生或近对生，革质，全缘略具波纹，无毛，叶脉在叶面明显凹。圆锥花序顶生或近枝顶腋生，密被星状毛。果近球形，黄褐色，假种皮肉质。花期春夏间，果期夏季。

Evergreen trees. Bark coarse, longitudinal crack. Leaves paripinnate, leaflets opposite or subopposite, leathery, both surfaces glabrous, margin entire,vines conspicuous impressed adaxially. Thyrses usually large, terminal or axillary near apex, densely stellate. Fruit usually yellowish brown, subglobose, with fleshy arillode. Fl. spring-summer, fr. summer.

树干 Trunk
摄影：吴林芳 Photo by: Wu Linfang

花枝 Flowering branch
摄影：卓书斌 Photo by: Zhuo Shubin

果枝 Fruiting branch
摄影：卓书斌 Photo by: Zhuo Shubin

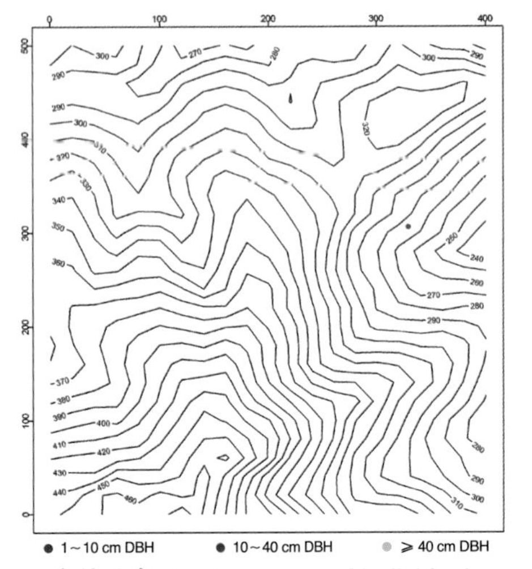

个体分布图 Distribution of individuals

径级分布表 DBH class

胸径等级 (Diameter class) (cm)	个体数 (No. of individuals in the plot)	比例 (Proportion) (%)
1～2	1	100.00
2～5	0	0.00
5～10	0	0.00
10～20	0	0.00
20～30	0	0.00
30～60	0	0.00
≥60	0	0.00

142 荔枝 lìzhī | Lychee

Litchi chinensis Sonn.

无患子科 | Sapindaceae

代码（SpCode）= LITCHI

个体数（Individual number/20 hm^2）= 1

最大胸径（Max DBH）= 1.5 cm

重要值排序（Importance value rank）= 188

常绿乔木。树皮灰褐色，通常光滑。偶数羽状复叶互生，小叶对生或近对生，革质，全缘略具波纹，无毛。圆锥花序顶生，阔大。果卵圆形至近球形，熟时暗红色或鲜红色，果皮有龟甲状裂纹，假种皮肉质。花期春夏季，果期夏季。

Evergreen trees. Bark grayish brown, usually smoothly. Leaves paripinnate, leaflets opposite or subopposite, leathery, both surfaces glabrous, margin entire ±wavy. Thyrses usually large, terminal. Fruit usually dark red to fresh red when mature, pericarp with tortoise-shell-like fissure, orbicular-ovate to subglobose, with fleshy arillode. Fl. spring-summer, fr. summer.

果枝 Fruiting branch
摄影：吴林芳 Photo by: Wu Linfang

枝叶 Branch and leaves
摄影：吴林芳 Photo by: Wu Linfang

叶背 Leaf abaxial surface
摄影：吴林芳 Photo by: Wu Linfang

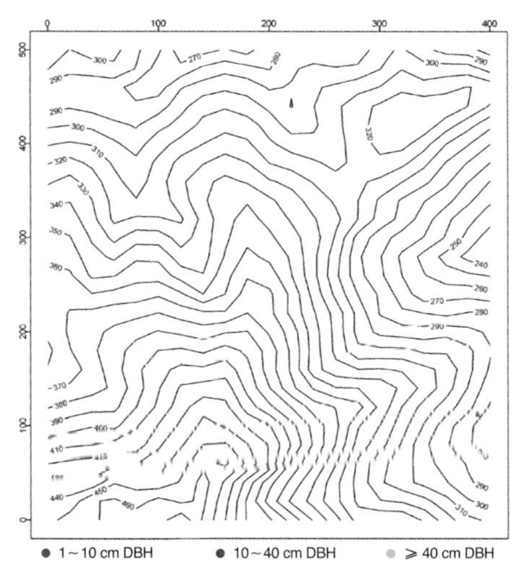

个体分布图 Distribution of individuals

径级分布表 DBH class

胸径等级 (Diameter class) (cm)	个体数 (No. of individuals in the plot)	比例 (Proportion) (%)
1～2	1	100.00
2～5	0	0.00
5～10	0	0.00
10～20	0	0.00
20～30	0	0.00
30～60	0	0.00
⩾60	0	0.00

143 褐叶柄果木　hèyèbǐngguǒmù | Brownleaf Mischocarpus

Mischocarpus pentapetalus (Roxb.) Radlk.
无患子科 | Sapindaceae

代码（SpCode）= MISPEN
个体数（Individual number/20 hm^2）= 1255
最大胸径（Max DBH）= 36.2 cm
重要值排序（Importance value rank）= 23

常绿乔木，高4～10m或更高。小枝短粗，幼时被毛。偶数羽状复叶，小叶对生或近对生，叶轴具槽，叶柄基部肿胀；小叶披针形或长披针形至长圆形，纸质或薄革质，无毛。花序常分枝，少数呈总状，单生叶腋或数个丛生于小枝近顶部。蒴果梨形或棒状。花期春季，果期夏季。

Evergreen trees, 4-10 m tall or more. Branches strong, short, only pubescent when young. Leaves paripinnate, leaflets opposite or subopposite, axis striate, petioles inflated at base, leaflets lanceolate or oblong-lanceolate to oblong, papery or thinly leathery, glabrous. Inflorescences often multibranched, rarely racemose, solitary and axillary or several fascicled near branch apices. Capsules pear-shaped or clavate. Fl. spring, fr. summer.

野外识别特征：

1. 叶为偶数羽状复叶，区别于橄榄的奇数复叶；
2. 小叶对生或近对生，叶轴具槽，叶柄基部肿胀；
3. 叶纸质或薄革质。

Key notes for identification:

1. Leaves paripinnate, distinguish from *Canarium album* odd-pinnate.
2. Leaflets opposite or subopposite, axis striate, petioles inflated at base.
3. Leaflets papery or thinly leathery.

花枝　Flowering branch
摄影：吴林芳　Photo by: Wu Linfang

叶背　Leaf abaxial surface
摄影：吴林芳　Photo by: Wu Linfang

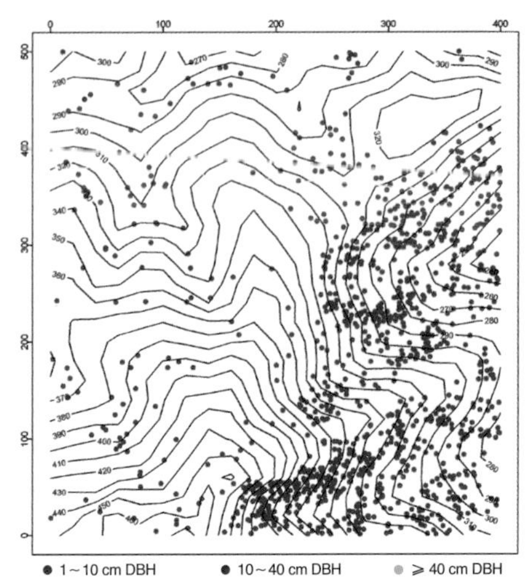

个体分布图 Distribution of individuals

径级分布表 DBH class

胸径等级 (Diameter class) (cm)	个体数 (No. of individuals in the plot)	比例 (Proportion) (%)
1～2	639	50.92
2～5	567	45.18
5～10	42	3.35
10～20	5	0.40
20～30	1	0.08
30～60	1	0.08
≥60	0	0.00

144 韶子 sháozǐ | Chinese Rambutan

Nephelium chryseum Bl.

无患子科 | Sapindaceae

代码（SpCode）= NEPCHR

个体数（Individual number/20 hm^2）= 153

最大胸径（Max DBH）= 42.5 cm

重要值排序（Importance value rank）= 54

常绿乔木，高10～20m或更高。小枝具槽，幼时被锈色茸毛。偶数羽状复叶，小叶对生或近对生，长圆形，通常10对，偶为2对，薄革质，叶背被毛。聚伞圆锥花序较大，顶生或近枝顶腋生。果红色，椭圆形，果皮具长约1cm的刺。花期春季，果期夏季。

Evergreen trees, 10-20 m tall or more. Branches striate, ferruginous pubescent when young. Leaves paripinnate, leaflets opposite or subopposite, oblong, leaflets 10 pairs, thinly leathery, pilose abaxially. Thyrses usually large, terminal or axillary near apex. Fruit red, ellipsoid, with spines ca. 1 cm. Fl. spring, fr. summer.

野外识别特征：

1. 小枝具槽，幼时被锈色茸毛；
2. 叶为偶数羽状复叶，小叶可达10对；
3. 小叶纸质或薄革质，近对生，叶背被毛。

Key notes for identification:

1. Branches striate, ferruginous pubescent when young.
2. Leaves paripinnate; leaflets usually 10 pairs.
3. Leaflets papery or thinly leathery, subopposite, pilose abaxially.

枝叶 Branch and leaves
摄影：吴林芳 Photo by: Wu Linfang

叶背 Leaf abaxial surface
摄影：吴林芳 Photo by: Wu Linfang

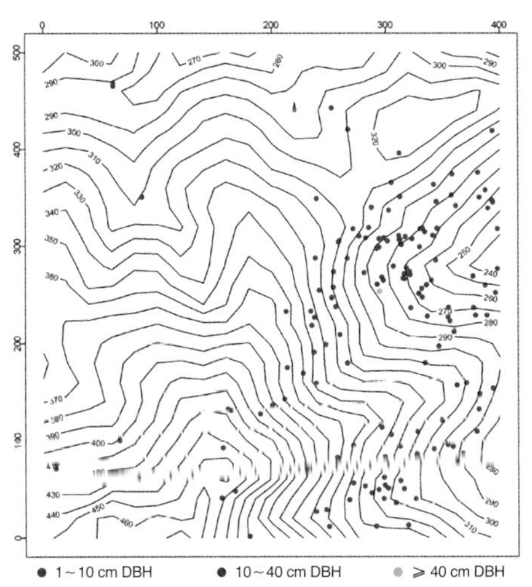

个体分布图 Distribution of individuals

径级分布表 DBH class

胸径等级 (Diameter class) (cm)	个体数 (No. of individuals in the plot)	比例 (Proportion) (%)
1～2	52	33.99
2～5	68	44.44
5～10	14	9.15
10～20	8	5.23
20～30	9	5.88
30～60	2	1.31
≥60	0	0.00

145 香皮树

xiāngpíshù | Ford's Meliosma

Meliosma fordii Hemsl.
清风藤科 | Sabiaceae

代码（SpCode）= MELFOR
个体数（Individual number/20 hm^2）= 3
最大胸径（Max DBH）= 17.9 cm
重要值排序（Importance value rank）= 147

常绿乔木，高达10m。树皮灰色。小枝、叶柄、叶背和花序被褐色茸毛。单叶，叶柄1.5～3.5cm长；倒披针形或狭倒卵形至狭椭圆形，近革质，全缘或顶部多少具齿。圆锥花顶生或近顶生。核果近球形或压扁的球形。花5～7月，果8～10月。

Evergreen trees, to 10 m tall. Bark gray. Branchlets, petioles, abaxial surface of leaf blade, and inflorescences brownish pubescent. Leaves simple, petiole 1.5-3.5 cm, leaf blade oblanceolate or narrowly obovate to narrowly elliptic, subleathery, margin entire or somewhat serrate near apex. Panicle terminal or axillary near apex. Drupe subglobose or compressed-globose. Fl. May-Jul., fr. Aug.-Oct..

野外识别特征：

1. 叶背疏被平伏状柔毛；
2. 侧脉每边11～20条；
3. 叶柄长1.5～3.5cm。

Key notes for identification:

1. Abaxial surface of leaves sparsely procumbent pubescent.
2. Leaf blade lateral veins 11-20 pairs.
3. Petiole 1.5-3.5 cm.

叶 Leaf
摄影：吴林芳 Photo by: Wu Linfang

叶背 Leaf abaxial surface
摄影：吴林芳 Photo by: Wu Linfang

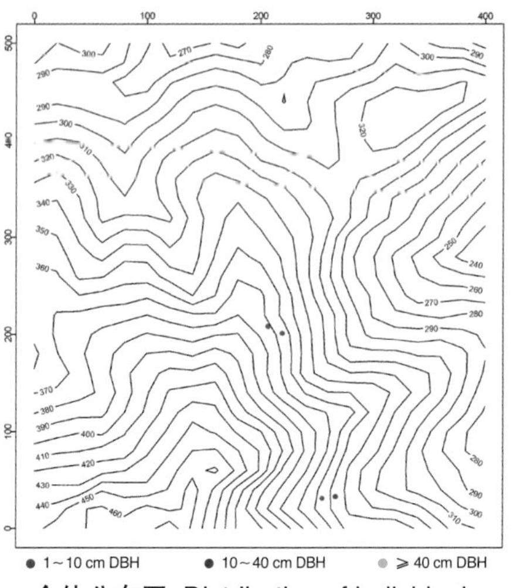

个体分布图 Distribution of individuals

径级分布表 DBH class

胸径等级 (Diameter class) (cm)	个体数 (No. of individuals in the plot)	比例 (Proportion) (%)
1～2	1	25.00
2～5	1	25.00
5～10	1	25.00
10～20	1	25.00
20～30	0	0.00
30～60	0	0.00
≥60	0	0.00

146 笔罗子 Biluozi | Stiff leaved Meliosma

Meliosma rigida Sieb. et Zucc.
清风藤科 | Sabiaceae

代码（SpCode）= MELRIG
个体数（Individual number/20 hm^2）= 116
最大胸径（Max DBH）= 23.3 cm
重要值排序（Importance value rank）= 47

常绿乔木，高7m。芽、嫩枝、叶背脉和花序被锈色茸毛或长茸毛。单叶，叶柄1.5～4cm；倒披针形或狭倒卵形，革质，叶背被锈色茸毛或密被长茸毛。边缘具粗齿，有时全缘。圆锥花序顶生，花梗三角形，3次分枝。核果球形。花期夏季，果期9～10月。

Evergreen trees, to 7 m tall. Buds, young branches, abaxial midveins, and inflorescences ferruginous lanuginous, pilose, or tomentose. Leaves simple, petiole 1.5-4 cm, leaf blade oblanceolate or narrowly obovate, leathery, abaxially ferruginous pubescent, or densely tomentose. Margin coarsely serrate, sometimes entire. Panicle terminal, axis triangular, branched 3 times. Drupe globose. Fl. summer, fr. Sep.-Oct..

野外识别特征：
1. 叶背密被锈色茸毛；
2. 侧脉每边9～18条；
3. 叶柄长1.5～4cm。

Key notes for identification:
1. Abaxial surface of leaves ferruginous pubescent or densely tomentose.
2. Leaf blade lateral veins 9-18 pairs.
3. Petiole 1.5-4 cm.

果枝 Fruiting branch
摄影：吴林芳 Photo by: Wu Linfang

枝叶 Branch and leaves
摄影：吴林芳 Photo by: Wu Linfang

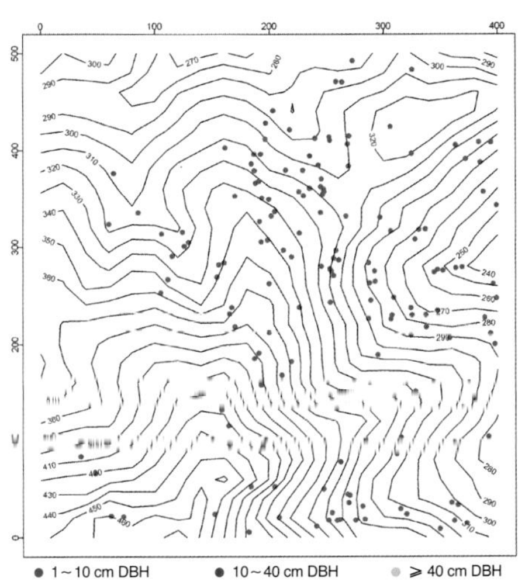

个体分布图 Distribution of individuals

径级分布表 DBH class

胸径等级 (Diameter class) (cm)	个体数 (No. of individuals in the plot)	比例 (Proportion) (%)
1～2	5	4.31
2～5	40	34.48
5～10	42	36.21
10～20	27	23.28
20～30	2	1.72
30～60	0	0.00
≥60	0	0.00

147 山羡花泡花树 shānxiànhuāpàohuāshù | Thorelii Meliosma

Meliosma thorelii Lecomte
清风藤科 | Sabiaceae

代码（SpCode）= MELTHO
个体数（Individual number/20 hm^2）= 5
最大胸径（Max DBH）= 11.3 cm
重要值排序（Importance value rank）= 123

常绿乔木，高14m。单叶，叶柄较短，约1.5～2cm；叶片倒披针状椭圆形、或倒披针形，革质，叶背无毛或疏被茸毛，全缘或具锐齿。圆锥花序顶生或近顶生。核果球形，略歪斜。花期夏季，果期10～11月。

Evergreen trees, to 14 m tall. Leaves simple, petiole 1.5-2 cm, leaf blade oblanceolate-elliptic or oblanceolate, leathery, abaxially glabrous or sparsely pubescent, margin entire or acutely serrulate. Panicle terminal or axillary on apical branches. Drupe globose, slightly oblique. Fl. summer, fr. Oct.-Nov..

野外识别特征：

1. 叶背无毛或疏被茸毛；
2. 侧脉15～28条每边；
3. 叶柄长1.5～2cm。

Key notes for identification:

1. Abaxial surface of leaves glabrous or sparsely pubescent.
2. Leaf blade lateral veins 15-28 pairs.
3. Petiole 1.5-2 cm.

叶 Leaf
摄影：吴林芳 Photo by: Wu Linfang

叶背 Leaf abaxial surface
摄影：吴林芳 Photo by: Wu Linfang

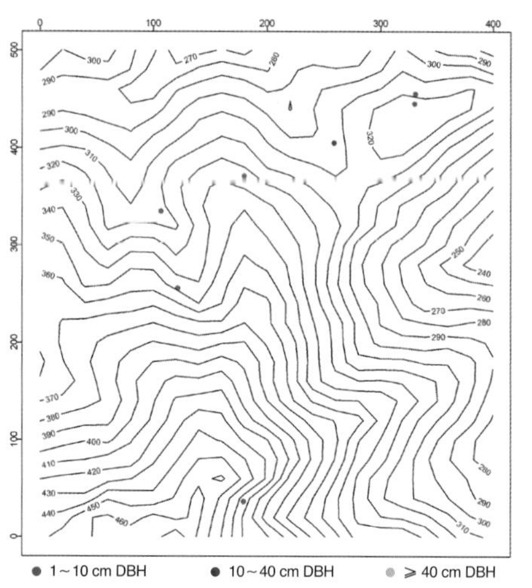

个体分布图 Distribution of individuals

径级分布表 DBH class

胸径等级 (Diameter class) (cm)	个体数 (No. of individuals in the plot)	比例 (Proportion) (%)
1～2	0	0.00
2～5	1	20.00
5～10	2	40.00
10～20	2	40.00
20～30	0	0.00
30～60	0	0.00
≥60	0	0.00

140 杧果 mángguǒ | Mango

Mangifera indica L.
漆树科 | Anacardiaceae

代码（SpCode）= MANIND
个体数（Individual number/20 hm^2）= 30
最大胸径（Max DBH）= 16.2 cm
重要值排序（Importance value rank）= 109

常绿乔木。小枝褐色，无毛。叶柄长2～6cm，基部彭大。叶革质，上面深绿，叶背亮绿，两面无毛，全缘，具波状，中脉两面凸起。圆锥花序顶生，无毛或多少具毛。核果长圆形或肾形，长5～10cm，宽3～4.5cm。花期3～4月，果期5～7月。

Evergreen trees. Branchlets brown, glabrous. Petiole 2-6 cm, inflated basally. Leaf blade leathery, deep green adaxially, light green abaxially, glabrous on both sides, margin entire, undulate, midrib prominent on both sides. Inflorescence paniculate, terminal, glabrous to tomentose-pilose. Drupe oblong to subreniform, 5-10 cm × 3-4.5 cm. Fl. Mar.-Apr., fr. May-Jul..

果 Fruit
摄影：吴林芳 Photo by: Wu Linfang

花序 Inflorescence
摄影：吴林芳 Photo by: Wu Linfang

叶 Leaf
摄影：吴林芳 Photo by: Wu Linfang

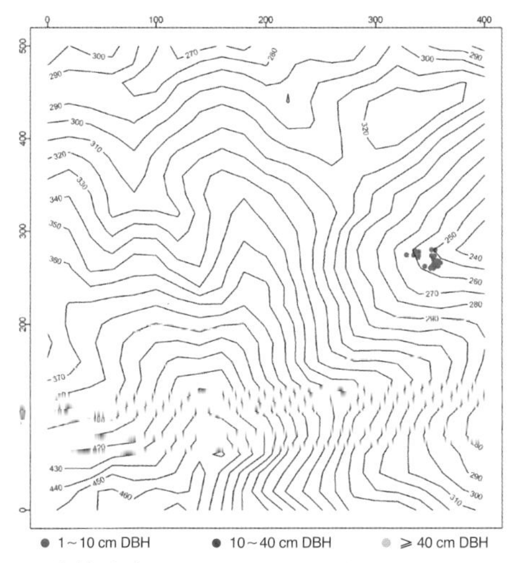

个体分布图 Distribution of individuals

径级分布表 DBH class

胸径等级 (Diameter class) (cm)	个体数 (No. of individuals in the plot)	比例 (Proportion) (%)
1～2	1	3.33
2～5	6	20.00
5～10	21	70.00
10～20	2	6.67
20～30	0	0.00
30～60	0	0.00
≥60	0	0.00

149 野漆

yěqī | Wax Tree

Toxicodendron succedaneum (Linn.) Kuntze
漆树科 | Anacardiaceae

代码（SpCode）= TOXSUC
个体数（Individual number/20 hm^2）= 6
最大胸径（Max DBH）= 7.8 cm
重要值排序（Importance value rank）= 134

落叶灌木或小乔木，高2～5m，很少达12m。全株无毛（别于木蜡树）。叶聚生枝顶，小叶7～13片，对生，纸质，长椭圆形或卵状披针形，全缘。圆锥花序腋生。核果歪斜，压扁，约6～8mm宽，黄色。花期春季，果期秋季。

Deciduous shrubs or small trees, 2-5 m tall seldom to 12 m. All parts glabrous (different from *T. sylvestre*). Leaves clustered on apex of branch, leaflets 7-13, opposite, papery, elliptic-oblong or ovate-lancceolate, margin entire. Panicles axillary. Drupe oblique, compressed, ca. 6-8 mm wide, yellow. Fl. spring, fr. autumn.

树干 Trunk
摄影：吴林芳 Photo by: Wu Linfang

果 Fruit
摄影：卓书斌 Photo by: Zhuo Shubin

花 Flower
摄影：卓书斌 Photo by: Zhuo Shubin

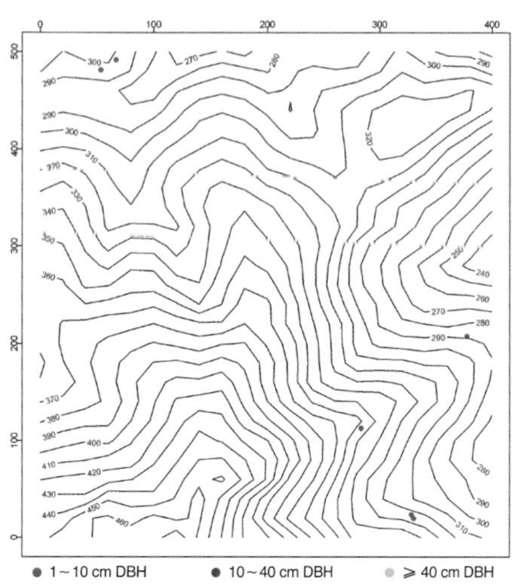

个体分布图 Distribution of individuals

径级分布表 DBH class

胸径等级 (Diameter class) (cm)	个体数 (No. of individuals in the plot)	比例 (Proportion) (%)
1～2	0	0.00
2～5	5	83.33
5～10	1	16.67
10～20	0	0.00
20～30	0	0.00
30～60	0	0.00
≥60	0	0.00

150 黄杞 huángqí | Roxburgh Engelhardtia

Engelhardia roxburghiana Wall.
胡桃科 | Juglandaceae

代码（SpCode）= ENGROX
个体数（Individual number/20 hm^2）= 654
最大胸径（Max DBH）= 95.0 cm
重要值排序（Importance value rank）= 4

常绿乔木，高达30m。树皮深褐色，粗糙具纵纹。幼枝及叶背等被黄褐色鳞秕。偶数羽状复叶，小叶3～5对，长圆形、长圆状椭圆形或卵形，全缘。单性花，花序顶生下垂。坚果具3叶状膜翅。花期5～7月，果期9～10月。

Evergreen trees to 30 m tall. Bark dark brown, coarse with longitudinal crack. Young branches and abaxial surfaces of leaves covered with brown peltate scales. Leaves even-pinnate, leaflets 3-5 pairs, oblong, oblong-elliptic or ovate, margin entire. Flowers unisexual, inflorescence terminal, pendulous. Fruit a 3-winged nutlet. Fl. May-Jul., fr. Sep.-Oct..

树干 Trunk
摄影：吴林芳 Photo by: Wu Linfang

叶 Leaf
摄影：吴林芳 Photo by: Wu Linfang

花 Flower
摄影：吴林芳 Photo by: Wu Linfang

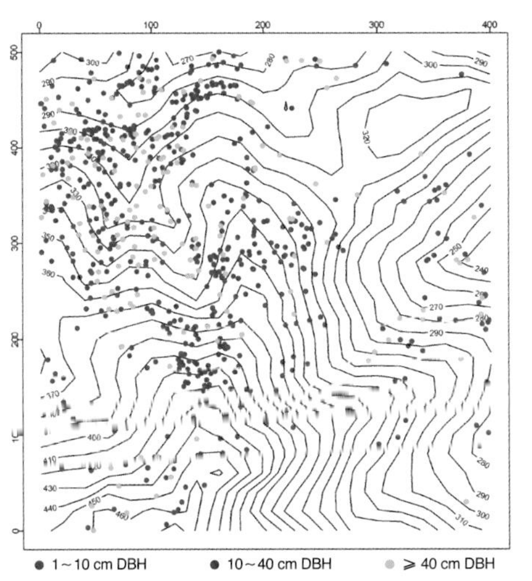

个体分布图 Distribution of individuals

径级分布表 DBH class

胸径等级 (Diameter class) (cm)	个体数 (No. of individuals in the plot)	比例 (Proportion) (%)
1～2	21	3.21
2～5	41	6.27
5～10	17	2.6
10～20	40	6.12
20～30	139	21.25
30～60	381	58.26
≥60	15	2.29

151 长刺楤木

chángcìcōngmù | Spine Aralia

Aralia spinifolia Merr.

五加科 | Araliaceae

代码（SpCode）= ARASPI

个体数（Individual number/20 hm^2）= 5

最大胸径（Max DBH）= 4.8 cm

重要值排序（Importance value rank）= 150

常绿灌木，高3m。花杂性同株。小枝密被刺毛并具直刺。叶二回羽状复叶，少数三回，具直刺。每羽具小叶5～9片，卵形至狭卵形，纸质至膜质，边缘具齿。伞状花序组成的圆锥花序顶生，被刚毛，具直刺。核果球形，径约5mm。花期8～10月，果期10～12月。

Evergreen shrubs, to 3 m tall. Andromonoecious. Branches with dense, flat prickles and slender setae. Leaves 2 (or 3)-pinnately compound, with straight prickles. Leaflets 5-9 per pinna, ovate to narrowly ovate, papery to membranous, margin serrate. Inflorescence a terminal panicle of umbels, setose, with prickles and bristles, lax. Drupe globose, ca. 5 mm in diam.. Fl. Aug.-Oct., fr. Oct.-Dec..

枝叶 Branch and leaves

摄影：吴林芳 Photo by: Wu Linfang

果 Fruit

摄影：吴林芳 Photo by: Zhuo Shubin

叶 Leaf

摄影：吴林芳 Photo by: Wu Linfang

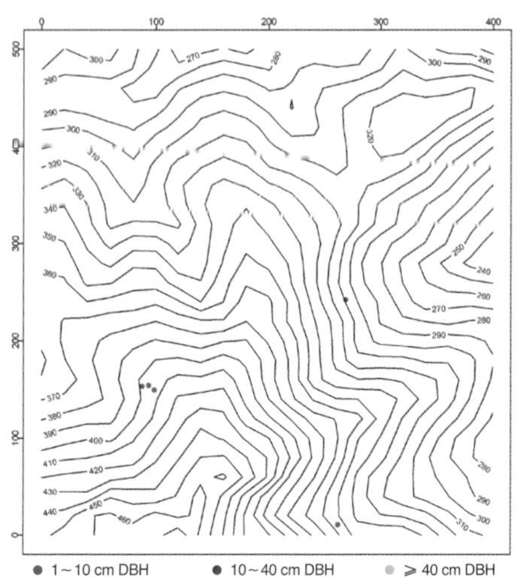

个体分布图 Distribution of individuals

径级分布表 DBH class

胸径等级 (Diameter class) (cm)	个体数 (No. of individuals in the plot)	比例 (Proportion) (%)
1～2	1	20.00
2～5	4	80.00
5～10	0	0.00
10～20	0	0.00
20～30	0	0.00
30～60	0	0.00
≥60	0	0.00

152 鹅掌柴（鸭脚木） Ezhangchai Ivy Tree

Schefflera heptaphylla (L.) Frodin
五加科 | Araliaceae

代码（SpCode）= SCHHEP
个体数（Individual number/20 hm^2）= 198
最大胸径（Max DBH）= 34.6 cm
重要值排序（Importance value rank）= 46

常绿乔木，高达15m。花杂性同株。掌状复叶，小叶6～9，有时达11片，纸质至革质，椭圆形至长椭圆形，幼时被毛后无毛。伞状花序组成的圆锥花序顶生，密被星状茸毛，后变无毛。核果球形，径约5mm。花期9～11月，果期12月至翌年2月。

Evergreen trees, 15 m tall. Andromonoecious. Leaves palmately compound, leaflets 6-9 (-11), elliptic to oblong-elliptic, papery to leathery, densely stellate pubescent when young, glabrescent. Inflorescence a terminal panicle of umbels, densely stellate tomentose, glabrescent. Drupe globose, ca. 5 mm in diam.. Fl. Sep.-Nov., fr. Dec.-Feb. of next year.

花 Flower
摄影：吴林芳 Photo by: Wu Linfang

果 Fruit
摄影：吴林芳 Photo by: Wu Linfang

叶 Leaf
摄影：吴林芳 Photo by: Wu Linfang

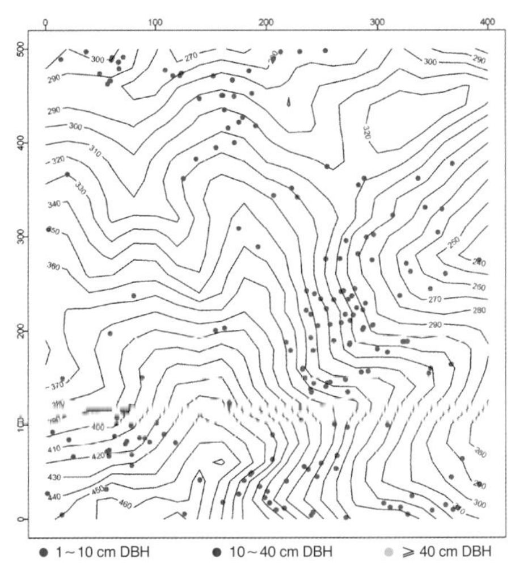

个体分布图 Distribution of individuals

径级分布表 DBH class

胸径等级 (Diameter class) (cm)	个体数 (No. of individuals in the plot)	比例 (Proportion) (%)
1～2	61	30.81
2～5	81	40.91
5～10	20	10.10
10～20	30	15.15
20～30	5	2.53
30～60	1	0.51
≥60	0	0.00

153 广东假木荷（红皮紫陵） guǎngdōngjiǎmùhé | Kwangtung Craibiodendron

Craibiodendron scleranthum var. *kwangtungense* (S. Y. Hu) Judd

杜鹃花科 | Ericaceae

代码（SpCode）= CRASCL

个体数（Individual number/20 hm²）= 3287

最大胸径（Max DBH）= 59.1 cm

重要值排序（Importance value rank）= 5

常绿小乔木，高5～12m。树皮深红褐色，不规则纵裂。小枝红褐色，具不明显皮孔。叶近椭圆形或披针形，革质，嫩叶红色。总状花序腋生。蒴果扁球形，5深沟。花期5～6月，果期7～8月。

Evergreen small trees, 5-12 m tall. Bark crimson-brown, irreqularly longitudinal crack. Branches red-brown, with unconspicuous lenticel. Leaf blade leathery, elliptic or lancceolate, young leaves reddish. Inflorescences axillary, racemose. Capsule depressed-globose, 5-loculicidal. Fl. May-Jun., fr. Jul.-Aug..

树干 Trunk
摄影：吴林芳 Photo by: Wu Linfang

果 Fruit
摄影：吴林芳 Photo by: Wu Linfang

叶 Leaf
摄影：吴林芳 Photo by: Wu Linfang

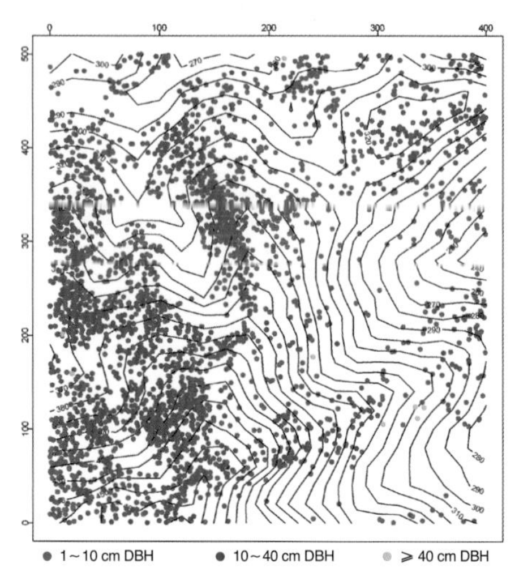

个体分布图 Distribution of individuals

径级分布表 DBH class

胸径等级 (Diameter class) (cm)	个体数 (No. of individuals in the plot)	比例 (Proportion) (%)
1～2	276	8.40
2～5	950	28.90
5～10	1007	30.64
10～20	807	24.55
20～30	200	6.08
30～60	47	1.43
≥60	0	0.00

[illegible] 吊钟花 diaozhonghua | Chinese New Year Flower

Enkianthus quinqueflorus Lour.
杜鹃花科 | Ericaceae

代码（SpCode）= RHOHEN
个体数（Individual number/20 hm^2）= 46
最大胸径（Max DBH）= 11.0 cm
重要值排序（Importance value rank）= 83

落叶灌木或小乔木，高1～3（～10）m。嫩枝无毛。叶革质，无毛，椭圆形、椭圆状披针形或倒卵状披针形，全缘或有时上部具波状细齿。伞状花序3～8朵花；花冠粉红或红色或白色，大体钟状。蒴果5棱，7～12mm。花期1～6月，果期3～9月。

Deciduous shrubs or small trees, 1-3 (-10) m tall. Twigs glabrous. Leaf blade leathery, elliptic, elliptic-lanceolate, or obovate-lanceolate, margin entire, sometimes sparsely sinuolate-serrulate towards apex. Inflorescence umbellate, 3-8-flowered, corolla pink, red, or white, broadly campanulate. Capsule 5-angled, 7-12 mm. Fl. Jan.-Jun., fr. Mar.-Sep..

花 Flower
摄影：吴林芳 Photo by: Wu Linfang

果 Fruit
摄影：吴林芳 Photo by: Wu Linfang

叶 Leaf
摄影：吴林芳 Photo by: Wu Linfang

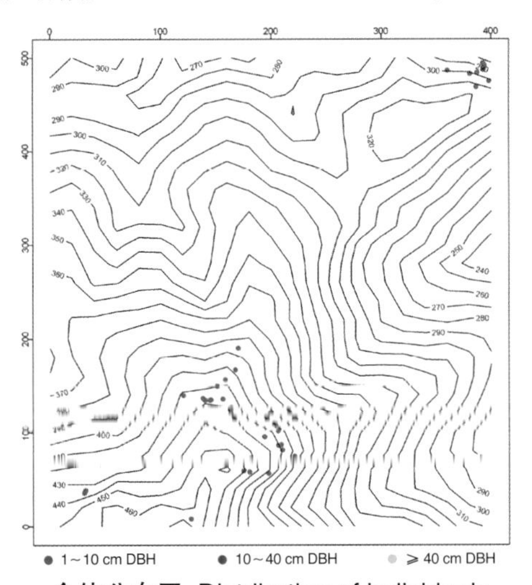

个体分布图 Distribution of individuals

径级分布表 DBH class

胸径等级 (Diameter class) (cm)	个体数 (No. of individuals in the plot)	比例 (Proportion) (%)
1～2	0	0.00
2～5	13	28.26
5～10	32	69.57
10～20	1	2.17
20～30	0	0.00
30～60	0	0.00
≥60	0	0.00

155 弯蒴杜鹃（罗浮杜鹃） wānshuòdùjuān | Loufu Rhododendron

Rhododendron henryi Hance
杜鹃花科 | Ericaceae

代码（SpCode）= RHO1HE
个体数（Individual number/20 hm^2）= 1404
最大胸径（Max DBH）= 23.9 cm
重要值排序（Importance value rank）= 18

常绿灌木或小乔木，高3～6m。小枝纤弱，灰褐色，多刚毛或腺状刚毛。叶卵状椭圆形或长披针形。伞状花序顶生，5～6朵花；花冠钟形漏斗状，紫色或粉红色。蒴果圆柱形，具中脉，略弯曲，长30～50mm。花期3～4月，果期9～12月。

Evergreen shrubs or small trees, 3-6 m tall. Branches slender, gray-brown, setose or glandular-setose. Leaf blade elliptic-ovate or oblong-lanceolate. Inflorescence terminal, umbellate, 5-6-flowered, corolla funnel-campanulate, purplish or pink. Capsule cylindric, with midrib, slightly curved, 30-50 mm. Fl. Mar.-Apr., fr. Sep.-Dec..

果 Fruit
摄影：吴林芳 Photo by: Wu Linfang

花 Flower
摄影：吴林芳 Photo by: Wu Linfang

叶 Leaf
摄影：吴林芳 Photo by: Wu Linfang

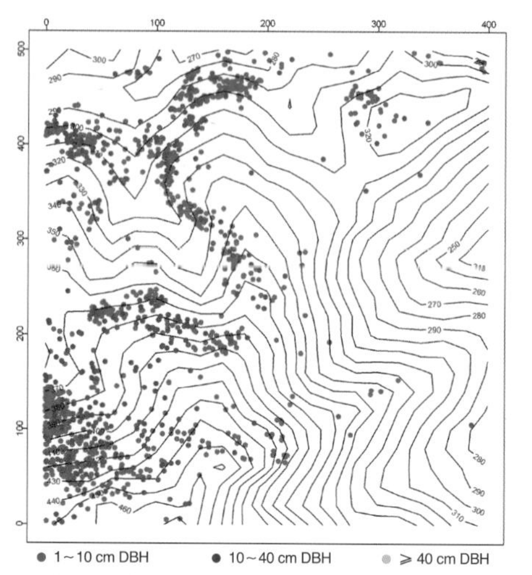

个体分布图 Distribution of individuals

径级分布表 DBH class

胸径等级 (Diameter class) (cm)	个体数 (No. of individuals in the plot)	比例 (Proportion) (%)
1～2	12	0.85
2～5	286	20.37
5～10	796	56.70
10～20	303	21.58
20～30	7	0.50
30～60	0	0.00
≥60	0	0.00

156 西施花（鹿角杜鹃）

Rhododendron latoucheae Franch.
杜鹃花科 | Ericaceae

代码（SpCode）= RHOLAT
个体数（Individual number/20 hm^2）= 9
最大胸径（Max DBH）= 11.3 cm
重要值排序（Importance value rank）= 126

常绿灌木或小乔木，高2～7m。单叶互生，聚生枝顶；小叶厚革质，狭椭圆状披针形、倒披针形或卵状椭圆形，叶缘略反卷，两面无毛。花序近顶生，1或2朵花，花冠漏斗状，粉红色或白色。蒴果圆柱形，长约40mm，宽4mm，花柱宿存。花期3～5月，果期9～10月。

Evergreen shrubs or small trees, 2-7 m tall. Leaves alternate, clustered at stem apex, leaf blade narrowly elliptic-lanceolate, oblanceolate or ovate-elliptic, thickly leathery, margin slightly revolute, both surfaces glabrous. Inflorescence subapical, 1(-2)-flowered, corolla funnelform, pink or white. Capsules cylindric, ca. 40 mm × 4 mm, style deciduous. Fl. Mar.-May, fr. Sep.-Oct..

花 Flower
摄影：吴林芳 Photo by: Wu Linfang

果 Fruit
摄影：吴林芳 Photo by: Wu Linfang

叶 Leaf
摄影：吴林芳 Photo by: Wu Linfang

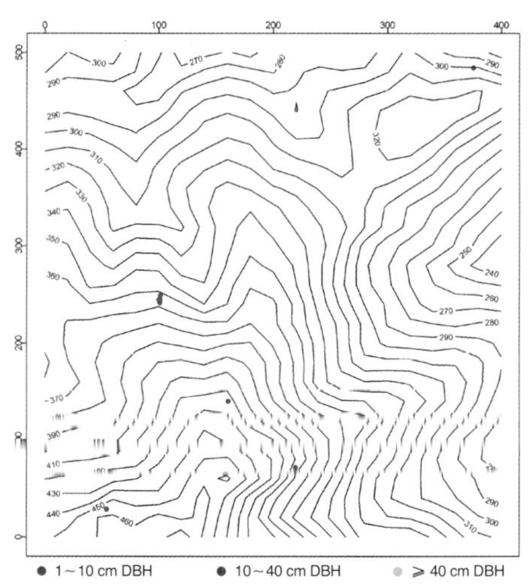

个体分布图 Distribution of individuals

径级分布表 DBH class

胸径等级 (Diameter class) (cm)	个体数 (No. of individuals in the plot)	比例 (Proportion) (%)
1～2	0	0.00
2～5	1	11.11
5～10	7	77.78
10～20	1	11.11
20～30	0	0.00
30～60	0	0.00
≥60	0	0.00

157 岭南杜鹃

lǐngnándùjuān | Lingnan Azalea

Rhododendron mariae Hance
杜鹃花科 | Ericaceae

代码（SpCode）= RHOMAR
个体数（Individual number/20 hm²）= 54
最大胸径（Max DBH）= 9.1 cm
重要值排序（Importance value rank）= 75

落叶灌木，高1～3（～7.5）m。老枝残留长毛；幼枝被红棕色糙伏毛。叶集于枝顶，革质，椭圆状披针形至椭圆状倒卵形，边缘略反卷，被糙伏毛。伞状花序具花7～16朵；花冠大体漏斗状，淡紫色。蒴果长卵形，长约9～14mm，宽3mm，密被红棕色糙伏毛。花期3～6月，果期7～11月。

Deciduous shrubs, 1-3(-7.5) m tall. Old branches with hair remains, young shoots coarsely red-brown strigose. Leaves clustered at stem apex, leaf blade leathery, elliptic-lanceolate to elliptic-obovate, margin slightly revolute, coarsely strigose. Inflorescence corymb, 7-16-flowered, corolla funnelform, lilac. Capsules long-ovoid, ca. 9-14 mm × 3 mm, densely coarsely red-brown-strigose. Fl. Mar.-Jun., fr. Jul.-Nov..

叶 Leaf
摄影：吴林芳 Photo by: Wu Linfang

花 Flower
摄影：吴林芳 Photo by: Wu Linfang

叶背 Leaf abaxial surface
摄影：吴林芳 Photo by: Wu Linfang

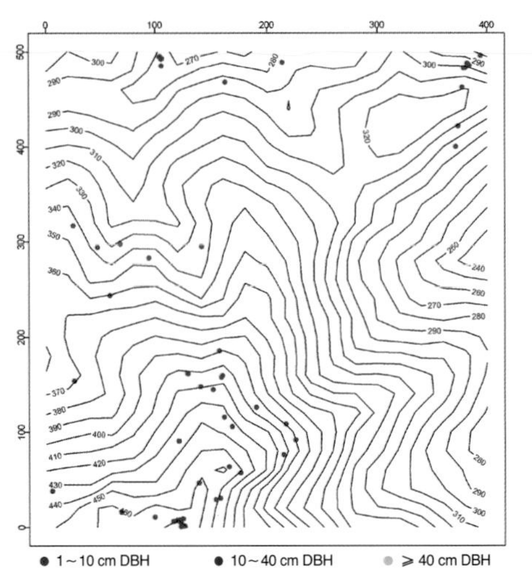

个体分布图 Distribution of individuals

径级分布表 DBH class

胸径等级 (Diameter class) (cm)	个体数 (No. of individuals in the plot)	比例 (Proportion) (%)
1～2	9	16.67
2～5	36	66.67
5～10	9	16.67
10～20	0	0.00
20～30	0	0.00
30～60	0	0.00
≥60	0	0.00

158 杜鹃（映山红） Dùjuān | Red Azalea

Rhododendron simsii Planch.
杜鹃花科 | Ericaceae

代码（SpCode）= RHOSIM
个体数（Individual number/20 hm^2）= 15
最大胸径（Max DBH）= 5.3 cm
重要值排序（Importance value rank）= 122

落叶灌木，高2（～5）m。分枝多而纤弱，密被亮棕褐色扁平糙毛。叶集生枝顶；卵形、椭圆状卵形或倒卵形至倒卵状披针形，边缘略反卷微具齿，纸质或革质。伞状花序2～3（～6）朵；花玫瑰色、鲜红或暗红色、白色或粉红色。蒴果卵形，密被糙伏毛，花萼宿存，花期4～5月，果期6～8月。

Deciduous shrubs, to 2 (-5) m tall. Branches many and fine, densely shiny brown appressed-setose, setae flat. Leaves clustered at stem apex, leaf blade ovate, elliptic-ovate or obovate to oblanceolate, margin slightly revolute, finely toothed, papery or leathery. Corymb 2-3 (-6)-flowered, corolla rose, bright to dark red, or white to rose-pink. Capsule ovoid, densely strigose, calyx persistent. Fl. Apr.-May, fr. Jun.-Aug..

叶背 Leaf abaxial surface
摄影：吴林芳 Photo by: Wu Linfang

花 Flower
摄影：吴林芳 Photo by: Wu Linfang

果枝 Fruiting branch
摄影：吴林芳 Photo by: Wu Linfang

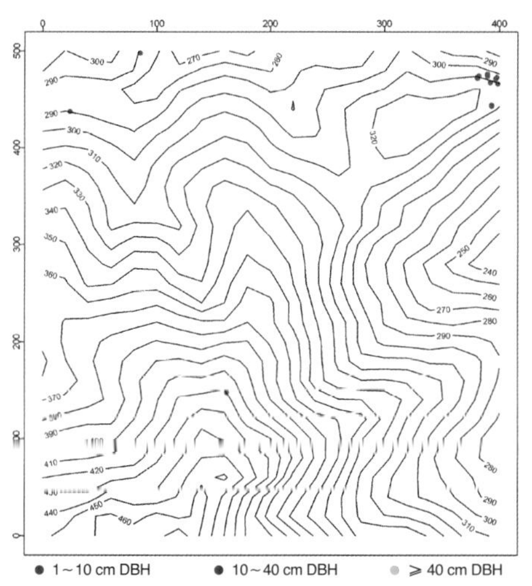

个体分布图 Distribution of individuals

径级分布表 DBH class

胸径等级 (Diameter class) (cm)	个体数 (No. of individuals in the plot)	比例 (Proportion) (%)
1～2	0	0.00
2～5	14	93.33
5～10	1	6.67
10～20	0	0.00
20～30	0	0.00
30～60	0	0.00
≥60	0	0.00

159 鼎湖杜鹃

dǐnghúdùjuān | Tingwu Azalea

Rhododendron tingwuense Tam
杜鹃花科 | Ericaceae

代码（SpCode）= RHOTIN
个体数（Individual number/20 hm^2）= 3
最大胸径（Max DBH）= 4.2 cm
重要值排序（Importance value rank）= 158

半常绿灌木，高2m。嫩枝纤弱近轮生，被糙伏毛。叶厚革质，椭圆形或倒卵状椭圆形，两面被糙伏毛。伞形花序常6～10朵，花冠狭漏斗状，淡紫色；雄蕊5，长出花冠。蒴果卵形，径约5mm，密被糙伏毛。花期3月，果期7～8月。

Semievergreen shrubs, to 2 m tall. Young shoots slender, subverticillate, strigose. Leaf blade thickly leathery, elliptic or obovate-elliptic, both surfaces coarsely strigose. Corymb usually 6-10-flowered, corolla narrowly funnelform, purplish, stamens 5, exserted from corolla. Capsules ovoid, ca. 5 mm, densely coarsely strigose. Fl. Mar., fr. Jul.-Aug..

花 Flower
摄影：刘莉 Photo by: Liu Li

果 Fruit
摄影：吴林芳 Photo by: Wu Linfang

叶 Leaf
摄影：吴林芳 Photo by: Wu Linfang

● 1～10 cm DBH ● 10～40 cm DBH ● ≥ 40 cm DBH

个体分布图 Distribution of individuals

径级分布表 DBH class

胸径等级 (Diameter class) (cm)	个体数 (No. of individuals in the plot)	比例 (Proportion) (%)
1～2	1	33.33
2～5	2	66.67
5～10	0	0.00
10～20	0	0.00
20～30	0	0.00
30～60	0	0.00
≥60	0	0.00

160 乌材 wūcái | Woollyflower Persimmon

Diospyros eriantha Champ. ex Benth.
柿科 | Ebenaceae

代码（SpCode）= DIOERI
个体数（Individual number/20 hm^2）= 249
最大胸径（Max DBH）= 27.4 cm
重要值排序（Importance value rank）= 45

常绿乔木或灌木，高2.5～15m。无顶芽。许多部分被锈色粗伏毛。叶互生，长圆状披针形，纸质。雄花1～3朵组成聚伞花序腋生；雌花单生叶腋。浆果卵形，紫色，长1.2～1.8cm，宽0.5～0.8cm，4室，被伏毛后略无毛。花期7～8月，果期10月至翌年2月。

Evergreen trees or shrubs, 2.5-15 m tall. Terminal buds absent. Many parts rusty strigose. Leaves alternate, leaf blade oblong-lanceolate, papery. Male flowers in axillary cymes, 1-3-flowered, female flowers solitary, axillary. Berries purple, ovoid, 1.2-1.8 cm × 0.5-0.8 cm, 4-locular, strigose, slightly glabrescent. Fl. Jul.-Aug., fr. Oct.-Feb. of next year.

花 Flower
摄影：吴林芳 Photo by: Wu Linfang

叶 Leaf
摄影：吴林芳 Photo by: Wu Linfang

果 Fruit
摄影：吴林芳 Photo by: Wu Linfang

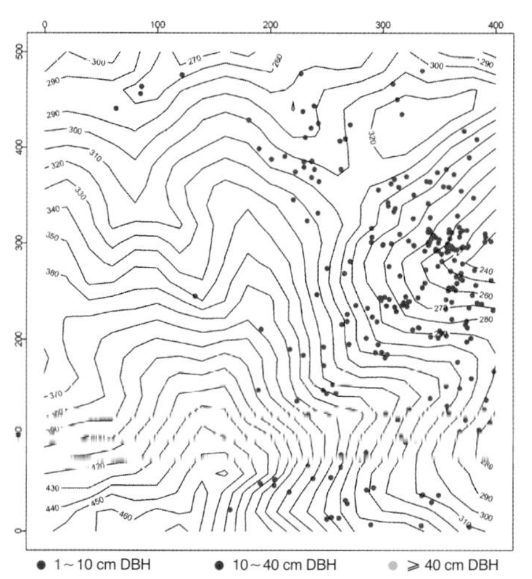

个体分布图 Distribution of individuals

径级分布表 DBH class

胸径等级 (Diameter class) (cm)	个体数 (No. of individuals in the plot)	比例 (Proportion) (%)
1～2	49	19.68
2～5	116	46.59
5～10	64	25.70
10～20	18	7.23
20～30	2	0.80
30～60	0	0.00
≥60	0	0.00

161 罗浮柿

luófúshì | Morris Persimmon

Diospyros morrisiana Hance
柿科 | Ebenaceae

代码（SpCode）= DIOMOR
个体数（Individual number/20 hm^2）= 291
最大胸径（Max DBH）= 25.7 cm
重要值排序（Importance value rank）= 38

落叶灌木或乔木，高3～20m。树皮薄片状剥落，黑色。除嫩枝、芽、花序被茸毛外其余无毛。叶薄革质，椭圆形，边缘有时略为波状。雄花序短小、聚伞状，下弯；雌花序单生；均腋生。浆果黄色，球形，径1.5～2.2(～2.9) cm，8室，无毛。花期5～6月，果期11月。

Deciduous shrubs or trees, 3-20 m tall. Bark peeling off in thin pieces, surface black. Young shoots, winter buds, inflorescences tomentose, otherwise glabrous. Leaf blade elliptic, thinly leathery, margin sometimes slightly undulate. Male flowers congested, cymose, nodding, female flowers solitary, axillary. Berries yellow, globose, 1.5-2.2(-2.9) cm in diam., 8-locular, glabrous. Fl. May-Jun., fr. Nov..

果 Fruit
摄影：吴林芳 Photo by: Wu Linfang

花 Flower
摄影：吴林芳 Photo by: Wu Linfang

叶 Leaf
摄影：吴林芳 Photo by: Wu Linfang

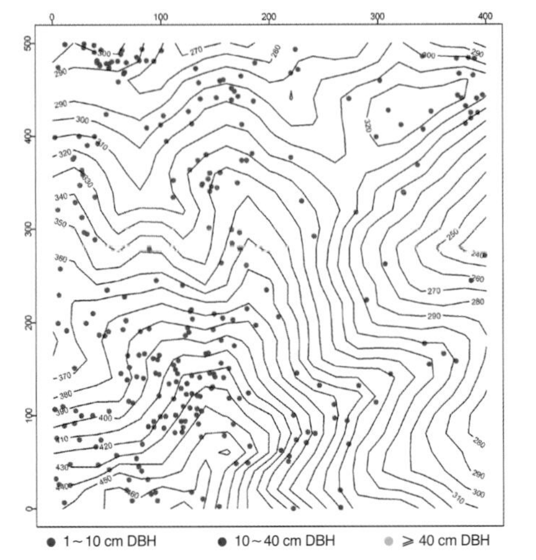

个体分布图 Distribution of individuals

径级分布表 DBH class

胸径等级 (Diameter class) (cm)	个体数 (No. of individuals in the plot)	比例 (Proportion) (%)
1～2	95	32.65
2～5	98	33.68
5～10	72	24.74
10～20	25	8.59
20～30	1	0.34
30～60	0	0.00
≥60	0	0.00

162 金叶树 [illegible]

Chrysophyllum lanceolatum BL.A.DC.
var. *stellatocarpon* P. Royen
山榄科 | Sapotaceae

代码（SpCode）= CHRLAN
个体数（Individual number/20 hm²）= 164
最大胸径（Max DBH）= 17.2 cm
重要值排序（Importance value rank）= 58

常绿乔木，高10～20m。小枝被黄色茸毛。单叶互生，叶柄2～7mm长，纸质，长圆形至长圆状披针形，少数倒卵形，侧脉多数而密集。花簇生叶腋。果褐色至紫黑色，近球形，具5棱，横向呈星状。花期5月，果期10月。

Evergreen trees, 10-20 m tall. Branchlets yellow pubescent. Leaves alternate, petiole 2-7 mm, leaf blade oblong to oblong-lanceolate, rarely obovate, papery, lateral veins many pairs closely spaced and slender. Flowers fascicled, axillary. Fruit brown to purplish black, subglobose, 5-ribbed into a stellate shape. Fl. May, fr. Oct..

野外识别特征：

1. 侧脉多数，纤细，整齐，几乎垂直于主脉；
2. 单叶互生，叶柄长2～7mm；
3. 嫩枝被黄色茸毛，绿色。

Key notes for identification:

1. Leaf blade lateral veins many pairs closely spaced and slender, orderliness, slightly perpendicular to midrib.
2. Leaves alternate, petiole 2-7 mm.
3. Branchlets yellow pubescent, green.

叶 Leaf
摄影：吴林芳 Photo by: Wu Linfang

叶背 Leaf abaxial surface
摄影：吴林芳 Photo by: Wu Linfang

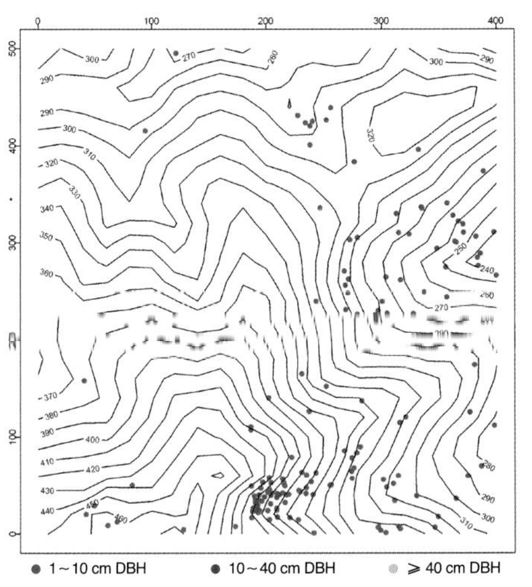

个体分布图 Distribution of individuals

径级分布表 DBH class

胸径等级 (Diameter class) (cm)	个体数 (No. of individuals in the plot)	比例 (Proportion) (%)
1～2	50	30.49
2～5	70	42.68
5～10	33	20.12
10～20	11	6.71
20～30	0	0.00
30～60	0	0.00
≥60	0	0.00

163 肉实树（水石梓） ròushíshù | Fleshy Nut Tree

Sarcosperma laurinum (Benth.) Hook. f.
肉实科 | Sarcopspermaceae

代码（SpCode）= SARLAU
个体数（Individual number/20 hm^2）= 1562
最大胸径（Max DBH）= 56.0 cm
重要值排序（Importance value rank）= 15

常绿乔木，高6～15（～26）m。树皮灰褐色，近平滑。小枝具棱，无毛。叶通常互生，有时对生，或在枝顶呈螺旋状着生；革质，通常倒卵形至倒披针形，两面无毛。总状或圆锥花序腋生。核果由绿转红最后成黑色，长圆形至椭圆形。花期8～9月，果期12月至翌年1月。

Evergreen trees, 6-15 (-26) m tall. Bark grayish brown, smooth. Branchlets angulate, glabrous. Leaves mostly alternate, some opposite, some whorled at end of branchlets, leaf blade usually obovate to oblanceolate, almost leathery, both surfaces glabrous. Racemes or panicles axillary. Corolla green, becoming pale yellow. Drupe green, becoming red and finally black, oblong to ellipsoid. Fl. Aug.-Sep., fr. Dec.-Jan. of next year.

花序 Inflorescence
摄影：吴林芳 Photo by: Wu Linfang

果 Fruit
摄影：吴林芳 Photo by: Wu Linfang

花蕾 bud
摄影：吴林芳 Photo by: Wu Linfang

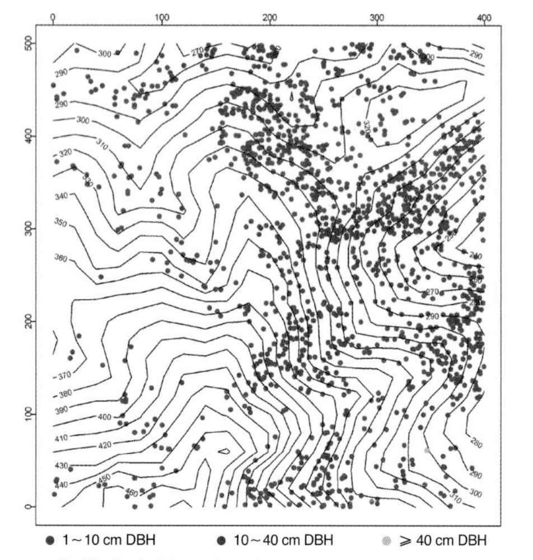

个体分布图 Distribution of individuals

径级分布表 DBH class

胸径等级 (Diameter class) (cm)	个体数 (No. of individuals in the plot)	比例 (Proportion) (%)
1～2	230	14.72
2～5	693	44.37
5～10	476	30.47
10～20	139	8.90
20～30	16	1.02
30～60	8	0.51
≥60	0	0.00

164 罗伞树 Luosanshu | Asiatic Ardisia

Ardisia quinquegona Bl.
紫金牛科 | Myrsinaceae

代码（SpCode）= ARDQUI
个体数（Individual number/20 hm^2）= 3696
最大胸径（Max DBH）= 17 cm
重要值排序（Importance value rank）= 10

常绿灌木，高2（~6）m。有时具地下茎。小枝具棱，幼时被锈色鳞片，后无毛，有纵纹。叶膜质，长圆形、椭圆形或倒披针形，侧脉多数。花序腋生，圆锥花序、聚伞花序或伞状花序。果核果状，扁球形，具5钝棱。花期3~7月，果期8月至翌年2月。

Evergreen shrubs, 2 (-6) m tall. Rarely rhizomatous. Branchlets angular, brown scaly, glabrescent, longitudinally ridged. Leaf blade oblong, elliptic, or oblanceolate, membranous, lateral veins numerous. Inflorescences axillary, paniculate, cymose, or subumbellate. Fruit drupaceous, spheroidicity, obtusely 5-angled. Fl. Mar.-Jul., fr. Aug.-Feb. of next year.

幼苗 Seedling
摄影：吴林芳 Photo by: Wu Linfang

花枝 Flowering branch
摄影：吴林芳 Photo by: Wu Linfang

果枝 Fruiting branch
摄影：吴林芳 Photo by: Wu Linfang

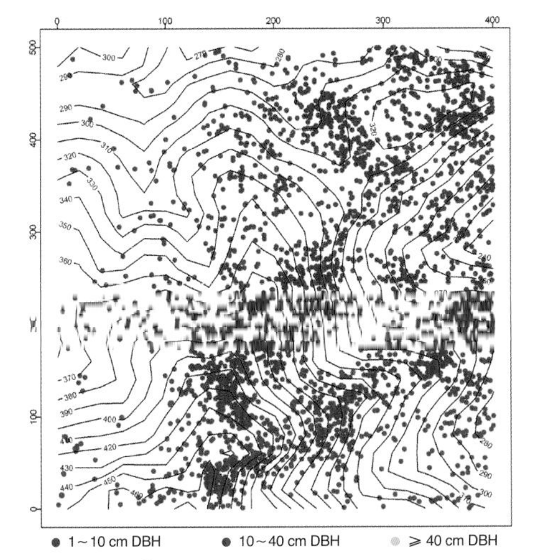

个体分布图 Distribution of individuals

径级分布表 DBH class

胸径等级 (Diameter class) (cm)	个体数 (No. of individuals in the plot)	比例 (Proportion) (%)
1~2	2111	57.12
2~5	1571	42.51
5~10	13	0.35
10~20	1	0.03
20~30	0	0.00
30~60	0	0.00
≥60	0	0.00

165 柳叶杜茎山

liǔyèdùjīngshān | Willow-leaf Maesa

Maesa salicifolia Walker
紫金牛科 | Myrsinaceae

代码（SpCode）= MAESAL
个体数（Individual number/20 hm^2）= 4
最大胸径（Max DBH）= 1.7 cm
重要值排序（Importance value rank）= 143

常绿灌木。小枝圆柱形，毛早脱落，具细纵纹。叶狭线状披针形，革质，多皱纹，脉间肿胀，边缘强烈反卷。总状或小圆锥花序腋生。果红色，球形或近卵形，径约4mm，具皱，宿存萼片几达宿存花柱。花期1～2月，果期9～11月。

Evergreen shrubs. Branchlets terete, early glabrescent, longitudinally striate. Leaf blade narrowly linear-lanceolate, leathery, rugose, subbullate, margin entire, strong revolute. Inflorescences axillary, racemose or paniculate. Fruit reddish, globose or subovoid, ca. 4 mm in diam., wrinkled, persistent calyx lobes nearly meeting style. Fl. Jan.-Feb., fr. Sep.-Nov..

果枝 Fruiting branch
摄影：吴林芳 Photo by: Wu Linfang

花 Flower
摄影：吴林芳 Photo by: Wu Linfang

叶 Leaf
摄影：吴林芳 Photo by: Wu Linfang

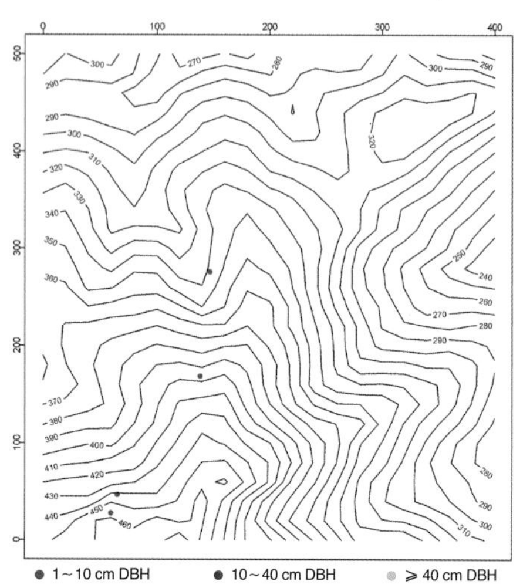

个体分布图 Distribution of individuals

径级分布表 DBH class

胸径等级 (Diameter class) (cm)	个体数 (No. of individuals in the plot)	比例 (Proportion) (%)
1～2	4	100.00
2～5	0	0.00
5～10	0	0.00
10～20	0	0.00
20～30	0	0.00
30～60	0	0.00
≥60	0	0.00

166 密花树 Mihuashu | Denseflower Myrsine

Myrsine seguinii H. Léveillé.
紫金牛科 | Myrsinaceae

果 Fruit
摄影：吴林芳 Photo by: Wu Linfang

代码（SpCode）= RAPNER
个体数（Individual number/20 hm^2）= 763
最大胸径（Max DBH）= 33.3 cm
重要值排序（Importance value rank）= 24

常绿灌木或乔木，高2~7（~12）m。小枝圆柱形，初时被红色茸毛，很快脱落，具皱纹，有时具白色皮孔。叶革质无毛，椭圆形至狭线状披针形，侧脉多数不明显。花3~10朵簇生。果灰绿色或紫黑色，球形或近卵圆形，径4~5mm。花期4~5月，果期10~12月。

Evergreen shrubs or trees, 2-7 (-12) m tall. Branchlets terete, white lenticellate, rugose, reddish puberulent, early glabrescent. Leaf blade elliptic to narrowly linear-oblanceolate, leathery, glabrous, lateral veins numerous, obscure. Inflorescences 3-10-flowered. Fruit grayish green or purplish black, globose or subovate, 4-5 mm in diam.. Fl. Apr.-May, fr. Oct.-Dec..

叶 Leaf
摄影：吴林芳 Photo by: Wu Linfang

花 Flower
摄影：吴林芳 Photo by: Wu Linfang

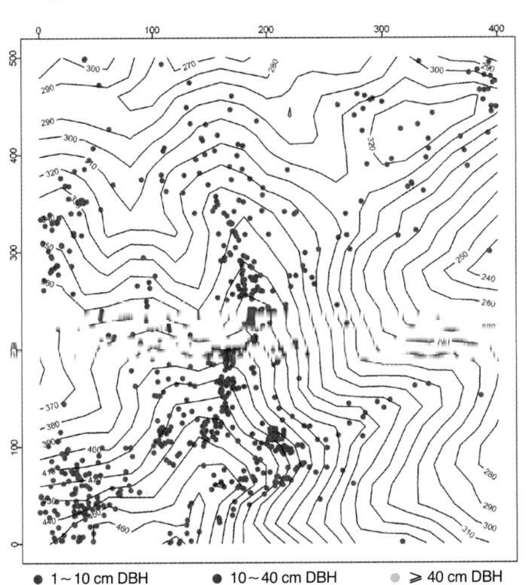

个体分布图 Distribution of individuals

径级分布表 DBH class

胸径等级 (Diameter class) (cm)	个体数 (No. of individuals in the plot)	比例 (Proportion) (%)
1~2	28	3.67
2~5	117	15.33
5~10	333	43.64
10~20	276	36.17
20~30	8	1.05
30~60	1	0.13
≥60	0	0.00

167 白花龙

báihuālóng | Faber Snowbell

Styrax faberi Perk.
安息香科 | Styracaceae

代码（SpCode）= STYFAB
个体数（Individual number/20 hm^2）= 4
最大胸径（Max DBH）= 12.3 cm
重要值排序（Importance value rank）= 113

落叶灌木或小乔木。小枝纤细，密披星状茸毛。叶互生；纸质，多少被棕色或灰色星毛，后逐渐无毛，边缘细锯齿。总状花序顶生，3～5朵花。果倒卵球形或近球形，密披灰色星毛，果皮光滑。花期5～10月。

Deciduous shrubs or small trees. Branchlets slender, densely stellate villose. Leaves alternate, leaf blade papery, sparsely brown to gray stellate pubescent to glabrescent, margin serrulate to remotely serrate. Racemes terminal, 3-5-flowered. Fruit obovoid or subglobose, densely gray stellate pubescent, exocarp smooth. Fr. May-Oct..

野外识别特征：

1. 落叶灌木或小乔木；
2. 叶背多少披棕色或灰色星毛，后逐渐无毛；
3. 叶纸质至薄革质。

Key notes for identification:

1. Deciduous shrubs or small tree.
2. Leaf blade abaxially sparsely brown to gray stellate pubescent to glabrescent.
3. Leaf blade papery to thinly leathery.

花 Flower
摄影：吴林芳 Photo by: Wu Linfang

叶 Leaf
摄影：吴林芳 Photo by: Wu Linfang

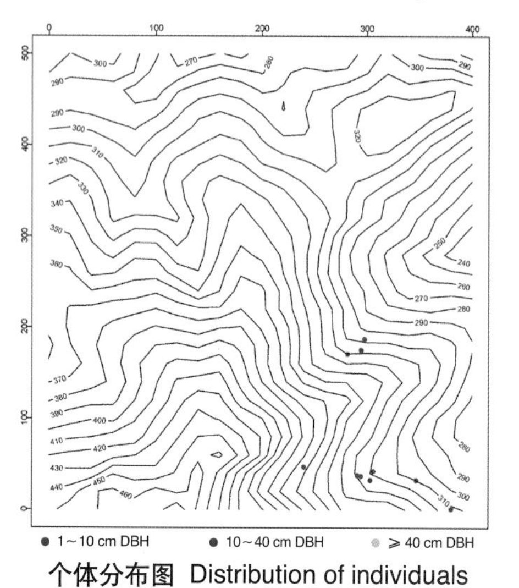

个体分布图 Distribution of individuals

径级分布表 DBH class

胸径等级 (Diameter class) (cm)	个体数 (No. of individuals in the plot)	比例 (Proportion) (%)
1～2	2	50.00
2～5	1	25.00
5～10	0	0.00
10～20	1	25.00
20～30	0	0.00
30～60	0	0.00
≥60	0	0.00

168 栓叶安息香

Styrax suberifolius Hook. et Arn.

安息香科 | Styracaceae

代码（SpCode）= STYSUB

个体数（Individual number/20 hm²）= 1

最大胸径（Max DBH）= 5.9 cm

重要值排序（Importance value rank）= 175

半常绿乔木，高4～20m。小枝被红褐色或灰褐色星状茸毛。叶互生，椭圆形、长圆形或椭圆状披针形，革质，叶背密被褐色茸毛。总状或圆锥花序顶生或腋生，多花。果卵球形，径1～1.8cm，密被灰色至褐色茸毛。花期3～5月，果期8～11月。

Semievergreen trees, 4-20 m tall. Branchlets red-brown to gray-brown stellate tomentose. Leaves alternate, leaf blade elliptic, oblong, or elliptic-lanceolate, leathery, abaxially densely brownish stellate tomentose. Inflorescences terminal or axillary, racemes or panicles, many-flowered. Fruit ovoid-globose, 1-1.8 cm in diam., densely gray to brown tomentose. Fl. Mar.-May, fr. Aug.-Nov..

野外识别特征：

1. 半常绿乔木；
2. 叶背密被褐色星状茸毛；
3. 叶革质。

Key notes for identification:

1. Semievergreen trees.
2. Leaf blade abaxially densely stellate tomentose.
3. Leaf blade leathery.

果枝 Fruiting branch

摄影：吴林芳 Photo by: Wu Linfang

叶背 Leaf abaxial surface

摄影：吴林芳 Photo by: Wu Linfang

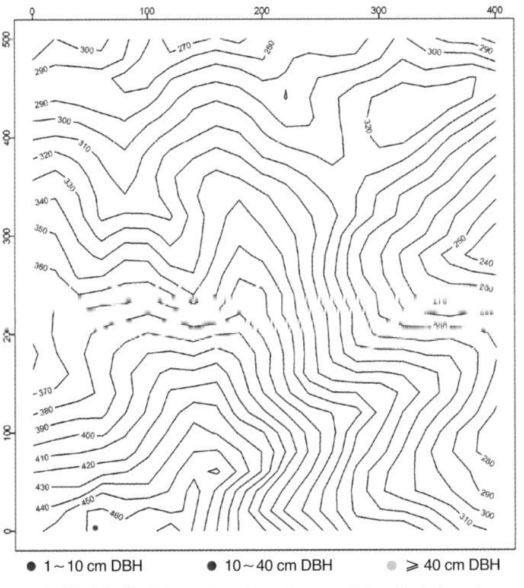

个体分布图 Distribution of individuals

径级分布表 DBH class

胸径等级 (Diameter class) (cm)	个体数 (No. of individuals in the plot)	比例 (Proportion) (%)
1～2	0	0.00
2～5	0	0.00
5～10	1	100.00
10～20	0	0.00
20～30	0	0.00
30～60	0	0.00
≥60	0	0.00

169 腺柄山矾 xiànbǐngshānfán | Glandularstipe Sweetleaf

Symplocos denopus Hance
山矾科 | Symplocaceae

代码（SpCode）= SYMADE
个体数（Individual number/20 hm^2）= 3
最大胸径（Max DBH）= 1.8 cm
重要值排序（Importance value rank）= 135

常绿灌木或小乔木。芽和嫩枝被茸毛但有时仅限于顶部，后无毛。叶柄长0.5～1.8cm，边缘密布腺齿；叶纸质，椭圆状卵形至卵形，叶缘密布腺齿。团伞花序腋生。核果圆柱形，长8～13mm，粗2.5～3mm，顶端具宿存萼片。花期11～12月，果期翌年7～8月。

Evergreen shrubs or small trees. Buds and young branchlets tomentose but sometimes only towards apex, glabrescent. Petiole 0.5-1.8 cm, marginal ridges densely glandular dentate, leaf blade elliptic-ovate to ovate, papery, margin densely glandular dentate. Inflorescences a glomerule, axillary. Drupes cylindrical, 8-13 mm × 2.5-3 mm, apex with persistent erect calyx lobes. Fl. Nov.-Dec., fr. Jul.- Aug. of next year.

野外识别特征：

1. 叶柄长0.5～1.8cm，边缘密布腺齿；
2. 叶纸质，叶缘密布腺齿；
3. 叶面脉间明显窿起。

Key notes for identification:

1. Petiole 0.5-1.8 cm, marginal ridges densely glandular dentate.
2. Leaf blade papery, margin densely glandular dentate.
3. Leaf blade subbullate among veins.

枝叶 Branch and leaves
摄影：吴林芳 Photo by: Wu Linfang

叶背 Leaf abaxial surface
摄影：吴林芳 Photo by: Wu Linfang

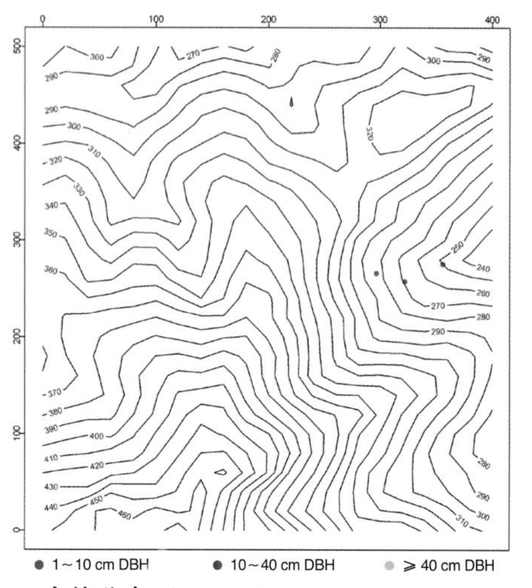

个体分布图 Distribution of individuals

径级分布表 DBH class

胸径等级 (Diameter class) (cm)	个体数 (No. of individuals in the plot)	比例 (Proportion) (%)
1～2	3	100.00
2～5	0	0.00
5～10	0	0.00
10～20	0	0.00
20～30	0	0.00
30～60	0	0.00
≥60	0	0.00

170 越南山矾 yuenanshanfan | Cochinchina Sweetleaf

Symplocos cochinchinensis (Lour.) S. Moore
山矾科 | Symplocaceae

代码（SpCode）= SYMCOC
个体数（Individual number/20 hm^2）= 1
最大胸径（Max DBH）= 15.0 cm
重要值排序（Importance value rank）= 184

常绿灌木或乔木。小枝粗壮。芽、嫩枝、叶柄和叶背中脉被红褐色茸毛。叶纸质，狭椭圆形、椭圆形或倒卵状椭圆形，近全缘或具细腺齿。穗状花序腋生，基部3～5分枝。核果灯泡形至近球形，径约5mm。花期8～9月，果期10～11月。

Evergreen shrubs or trees. Branchlets stocky. Bud, twigs, petiole and midrib of abaxially reddish brown tomentose. Leaf blade narrowly elliptic, elliptic, or obovate-elliptic, papery, margin subentire to glandular dentate. Spikes axillary, 3-5-branched from base. Drupes ampulliform to subglobose ca. 5mm. Fl. Aug.-Sep., fr. Oct.-Nov..

野外识别特征：

1. 小枝粗壮；
2. 嫩枝、叶柄和叶背中脉被红褐色，茸毛；
3. 叶纸质至革质，近全缘或具细腺齿。

Key notes for identification:

1. Branchlets stocky.
2. Twigs, petiole and midrib of abaxially reddish brown tomentose.
3. Leaf blade papery to leathery, margin subentire to glandular dentate.

果 Fruit
摄影：吴林芳 Photo by: Wu Linfang

叶 Leaf
摄影：吴林芳 Photo by: Wu Linfang

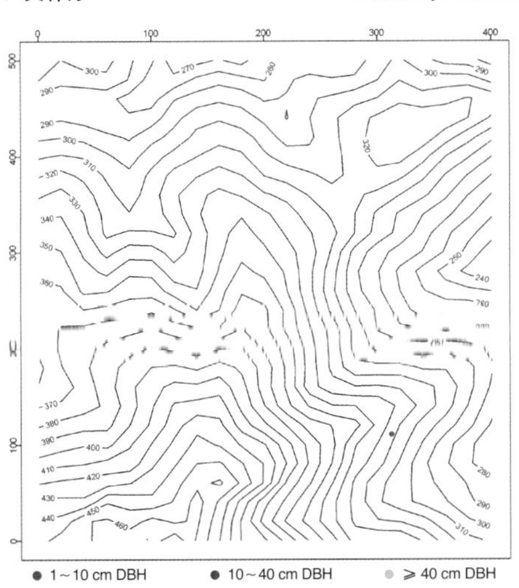

个体分布图 Distribution of individuals

径级分布表 DBH class

胸径等级 (Diameter class) (cm)	个体数 (No. of individuals in the plot)	比例 (Proportion) (%)
1～2	0	0.00
2～5	1	100.00
5～10	0	0.00
10～20	0	0.00
20～30	0	0.00
30～60	0	0.00
≥60	0	0.00

171 光叶山矾

guānyèshānfán | Smoothleaf Sweetleaf

Symplocos lancifolia Sieb. et Zucc.
山矾科 | Symplocaceae

代码（SpCode）= SYMLAN
个体数（Individual number/20 hm^2）= 10
最大胸径（Max DBH）= 13.7 cm
重要值排序（Importance value rank）= 106

常绿灌木或乔木，高达20m。芽、嫩枝和花序轴被贴生长毛。小枝黑褐色，变无毛。叶近膜质至纸质，卵形、椭圆形、狭卵形或狭椭圆形，叶缘具齿。穗状花序腋生，基部不分枝。核果椭圆形至近球形，具宿存花萼。花期3～11月，果期6～12月。

Evergreen shrubs or trees, to 20 m tall. Buds, young branchlets, and inflorescence axes appressed to patently hairy. Branchlets dark brown, glabrescent. Leaf blade ovate, elliptic, narrowly ovate, or narrowly elliptic, submembranous to papery, margin finely crenate to dentate. Spikes axillary, no branched. Drupes ellipsoid to subglobose, apex with persistent calyx lobes. Fl. Mar.-Nov., fr. Jun.-Dec..

野外识别特征：

1. 芽、嫩枝和花序轴被贴生长毛；
2. 叶近膜质至纸质，叶缘具齿；
3. 叶先端尾状渐尖。

Key notes for identification:

1. Buds, young branchlets, and inflorescence axes appressed to patently hairy.
2. Leaf blade submembranous to papery, margin finely crenate to dentate.
3. Leaf blade apex caudate-acuminate.

果枝 Fruiting branch
摄影：吴林芳 Photo by: Wu Linfang

花 Flower
摄影：吴林芳 Photo by: Wu Linfang

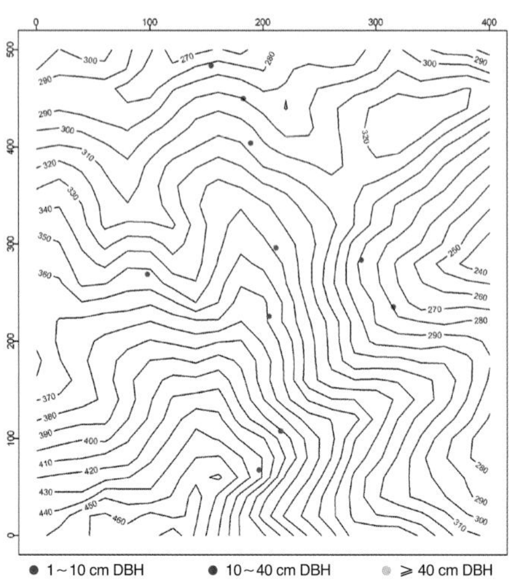

个体分布图 Distribution of individuals

径级分布表 DBH class

胸径等级 (Diameter class) (cm)	个体数 (No. of individuals in the plot)	比例 (Proportion) (%)
1～2	1	10.00
2～5	2	20.00
5～10	2	20.00
10～20	5	50.00
20～30	0	0.00
30～60	0	0.00
≥60	0	0.00

172 微毛山矾 wēimáoshānfán | Puberulous Sweetleaf

Symplocos wikstroemiifolia Hayata

山矾科 | Symplocaceae

代码（SpCode）= SYMWIK

个体数（Individual number/20 hm²）= 92

最大胸径（Max DBH）= 38.8 cm

重要值排序（Importance value rank）= 56

常绿灌木或乔木。幼枝、叶柄、叶背被紧贴的细毛。叶聚生枝顶，狭椭圆形、椭圆形、狭倒卵形或倒卵形，纸质至薄革质，全缘或具波状齿。花序总状，雄花、两性花大体异株。核果卵形，熟时黑色，宿存萼片直立。花期3月，果期10月。

Evergreen shrubs or trees. Young branchlets, petioles, and leaf blades abaxially with minute appressed hairs. Leaves only towards end of twigs, leaf blade narrowly elliptic, elliptic, narrowly obovate, or obovate, papery to thinly leathery, margin entire or sinuolate-dentate. Flowers male or bisexual, probably androdioecious, raceme. Drupes ovoid, dark, apex with persistent erect calyx lobes. Fl. Mar., fr. Oct..

野外识别特征：

1. 幼枝、叶柄、叶背被紧贴的细毛；
2. 叶聚生枝顶；
3. 叶纸质至薄革质，全缘或具波状齿。

Key notes for identification:

1. Young branchlets, petioles, and leaves abaxially with minute appressed hairs.
2. Leaves only towards end of twigs.
3. Leaf blade papery to thinly leathery, margin entire or sinuolate-dentate.

花 Flower

摄影：吴林芳 Photo by: Wu Linfang

叶背 Leaf abaxial surface

摄影：吴林芳 Photo by: Wu Linfang

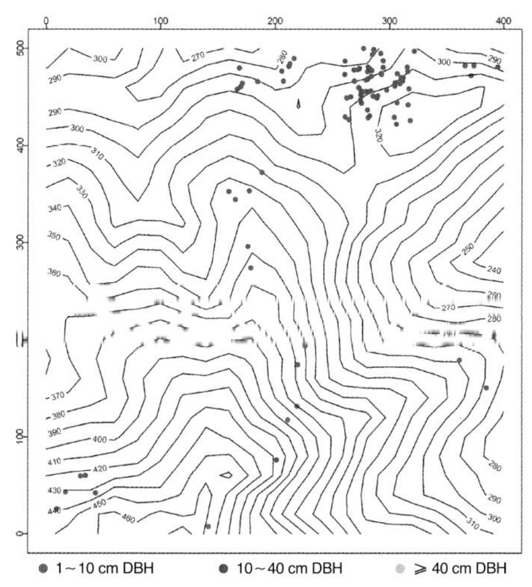

个体分布图 Distribution of individuals

径级分布表 DBH class

胸径等级 (Diameter class) (cm)	个体数 (No. of individuals in the plot)	比例 (Proportion) (%)
1～2	0	0.00
2～5	2	2.17
5～10	4	4.35
10～20	44	47.83
20～30	38	41.30
30～60	4	4.35
≥60	0	0.00

173 异株木犀榄（云南木犀榄） yìzhūmùxīlǎn | Dioecious Olive

Olea tsoongii (Merr.) P.S.Green
木犀科 | Oleaceae

代码（SpCode）= CHIRAM
个体数（Individual number/ 20 hm^2）= 5
最大胸径（Max DBH）= 10.2 cm
重要值排序（Importance value rank）= 170

常绿灌木或小乔木，高3～15m。杂性异株。小枝圆柱形，略被毛后无毛。叶对生，革质，全缘或上部具细齿。圆锥状花序腋生，被毛后变无毛。核果紫黑色，长6～13mm，宽3～9mm。花期2～11月，果期5～12月。

Evergreen shrubs or small trees, 3-15 m. Polygamodioecious. Branchlets terete, finely pubescent to glabrescent. Leaves opposite, leathery, margin entire or finely serrate. Inflorescences axillary, paniculate, puberulent to glabrescent. Drupe purple-black, 6-13 mm × 3-9 mm. Fl. Feb.-Nov., fr. May-Dec..

野外识别特征：
1. 常绿灌木或小乔木；
2. 叶对生，革质；
3. 叶全缘或上部具细齿。

Key notes for identification:
1. Evergreen shrubs or small trees.
2. Leaves opposite, leathery.
3. Leaf blade margin entire or apically finely serrate.

叶 Leaf
摄影：吴林芳 Photo by: Wu Linfang

叶背 Leaf abaxial surface
摄影：吴林芳 Photo by: Wu Linfang

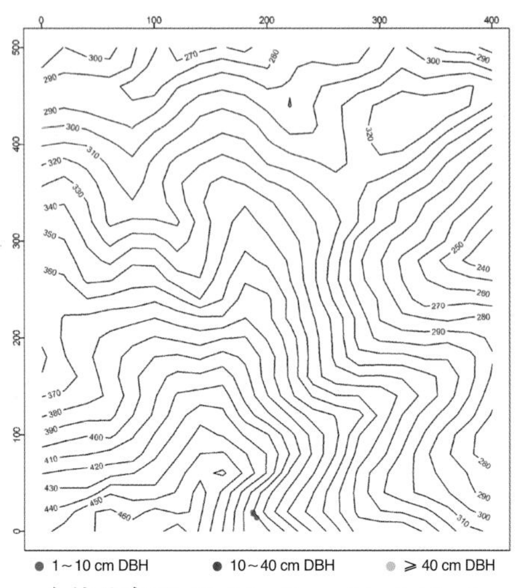

个体分布图 Distribution of individuals

径级分布表 DBH class

胸径等级 (Diameter class) (cm)	个体数 (No. of individuals in the plot)	比例 (Proportion) (%)
1～2	0	0.00
2～5	3	60.00
5～10	1	20.00
10～20	1	20.00
20～30	0	0.00
30～60	0	0.00
≥60	0	0.00

174 香楠（光叶山黄皮） xiāngnán | Mountain Wanpi

Aidia canthioides (Champ. ex Benth.) Masamune
茜草科 | Rubiaceae

代码（SpCode）= AIDCAN
个体数（Individual number/20 hm^2）= 5446
最大胸径（Max DBH）= 31.7 cm
重要值排序（Importance value rank）= 6

常绿灌木或乔木，高1～12 m，无刺，小枝无毛。单叶对生，纸质或薄革质；长椭圆形至长披针形，无毛。聚伞花序伞状，腋生；花萼背被锈色长毛，子房2室。浆果球形，具6～7粒种子。花期4～9月，果期7～12月。

Evergreen shrubs or trees, 1-12m tall. Unarmed. Branches glabrous. Leaves opposite, papery or thinly leathery, leaf blade oblong-elliptic to oblong-lanceolate, glabrous. Cymes umbel-like, axillary, calyx covered rusty hairs abaxially, ovary 2-locular. Berry globose with 6-7-seeds. Fl. Apr.-Sep., fr. Jul.-Dec..

花 Flower
摄影：吴林芳 Photo by: Wu Linfang

叶 Leaf abaxial surface
摄影：吴林芳 Photo by: Wu Linfang

果枝 Fruiting branch
摄影：吴林芳 Photo by: Wu Linfang

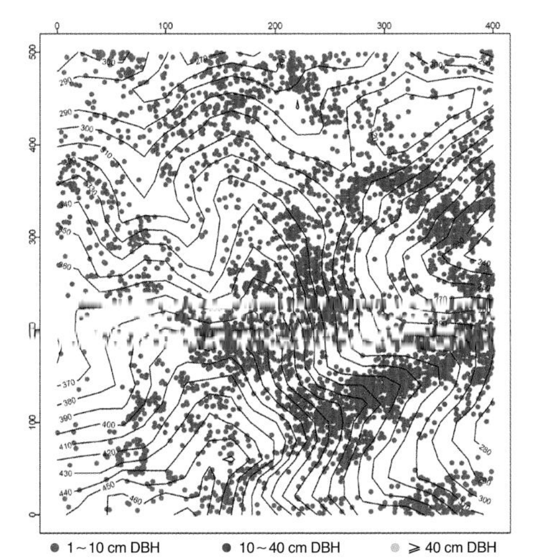

个体分布图 Distribution of individuals

径级分布表 DBH class

胸径等级 (Diameter class) (cm)	个体数 (No. of individuals in the plot)	比例 (Proportion) (%)
1～2	3427	62.93
2～5	1536	28.20
5～10	444	8.15
10～20	37	0.68
20～30	1	0.02
30～60	1	0.02
≥60	0	0.00

175 鱼骨木 yúgǔmù | Butulang Canthium

Canthium dicoccum (Gaertn.) Merr.
茜草科 | Rubiaceae

代码（SpCode）= CANDIC
个体数（Individual number/20 hm^2）= 582
最大胸径（Max DBH）= 37.6 cm
重要值排序（Importance value rank）= 25

常绿灌木或乔木，高达15m。无刺。小枝黑褐色，无毛。叶对生，革质，卵形或椭圆形至卵状披针形。聚伞花序腋生，松散而多花，花萼顶截平或具5浅齿。核果倒卵形或倒卵状椭圆体，径6～8mm。花果期3～8月。

Evergreen shrubs or trees, to 15 m tall. Spineless. Branchlets dark brown, glabrous. Leaves opposite, leathery, leaf blade ovate or elliptic or ovate-lanceolate. Cymes axillary, loose and often many flowered, calyx apex truncate or inconspicuously 5-toothed. Drupes obovate or obovate-ellipsoidal, 6-8 mm in diam.. Fl. and fr. Mar.-Aug..

野外识别特征：

1. 树皮灰褐色，无刺，多树瘤；
2. 小枝初时压扁或具棱，后圆柱形，无毛；
3. 叶对生，排列成假二回羽状复叶。

Key notes for identification:

1. Bark gray-brown, spineless, with obvious burl.
2. Branchlets slightly compressed or angular when young,then terete, glabrous.
3. Leaves opposite, pseudo-bipinnate.

叶 Leaf
摄影：吴林芳 Photo by: Wu Linfang

花 Flower
摄影：吴林芳 Photo by: Wu Linfang

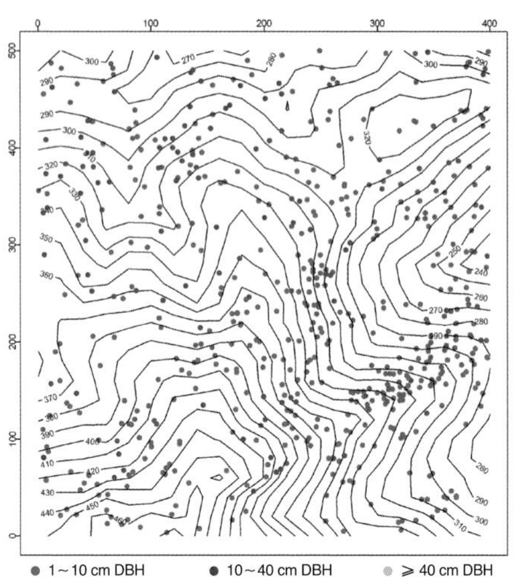

个体分布图 Distribution of individuals

径级分布表 DBH class

胸径等级 (Diameter class) (cm)	个体数 (No. of individuals in the plot)	比例 (Proportion) (%)
1～2	255	42.15
2～5	193	31.90
5～10	47	7.77
10～20	84	13.88
20～30	24	3.97
30～60	2	0.33
≥60	0	0.00

176 猪肚木 zhudumu | Bristly Canthium

Canthium horridum Bl.
茜草科 | Rubiaceae

代码（SpCode）= CANHOR
个体数（Individual number/20 hm²）= 23
最大胸径（Max DBH）= 7.8 cm
重要值排序（Importance value rank）= 95

常绿灌木，高2～3m。具腋生直刺。幼枝微被毛。叶对生，纸质，卵形或椭圆形，侧脉2～3对。花小，单个或多个簇生叶腋，花萼顶不明显的波状齿。核果球形或孪生，径1～2cm。花期4～6月，果期7～11月。

Evergreen shrubs, 2-3m tall. Armed with straight axillary thorns. Young branchlets slightly pubescent. Leaves opposite, papery, leaf blade ovate or elliptic, lateral veins 2-3 on eachside of midvein. Flowers small, solitary or clustered in axils of leaves, calyx apes with minute, undulate teeth. Drupes globular or didymous, 1-2 cm in diam.. Fl. Apr.-Jun., fr. Jul.-Nov..

野外识别特征：
1. 树干具腋生直刺；
2. 枝纤细，圆柱形，不具短侧枝；
3. 核果球形或孪生，不具沟槽。

Key notes for identification:
1. Trunk armed eith straight axillary thorns.
2. Branchlets slender, terete, no very short lateral branches.
3. Drupes globular or didymous, no grooves on surface.

叶 Leaf
摄影：吴林芳 Photo by: Wu Linfang

树干及果 Trunk & Fruit
摄影：吴林芳 Photo by: Wu Linfang

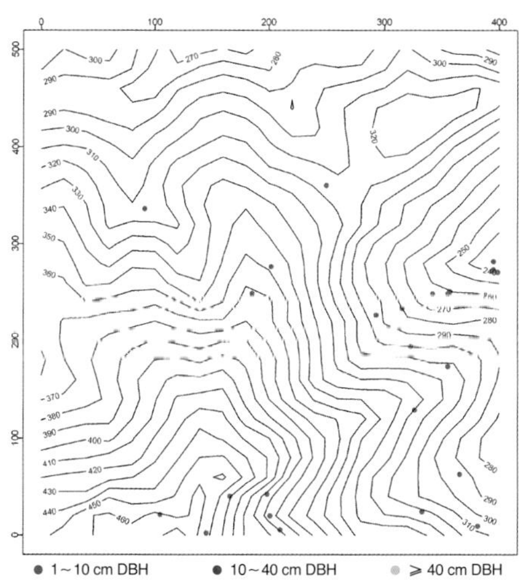

个体分布图 Distribution of individuals

径级分布表 DBH class

胸径等级 (Diameter class) (cm)	个体数 (No. of individuals in the plot)	比例 (Proportion) (%)
1～2	12	52.17
2～5	10	43.48
5～10	1	4.35
10～20	0	0.00
20～30	0	0.00
30～60	0	0.00
≥60	0	0.00

177 山石榴

shānshíliú | Spiny Randia

Catunaregam spinosa (Thunb.) Tirveng
茜草科 | Rubiaceae

代码（SpCode）= CATSPI
个体数（Individual number/20 hm²）= 8
最大胸径（Max DBH）= 15.0 cm
重要值排序（Importance value rank）= 125

常绿灌木或小乔木，高1～10m。多分枝，具腋对生的粗刺。叶对生或簇生短侧枝。叶纸质或近革质，倒卵形或长倒卵形，侧脉4～7对。花单个或2～3朵簇生短侧枝顶端，花萼筒5裂。浆果球形，具沟槽，径约2～4cm。花期3～6月，果期5月至翌年1月。

Evergreen shrubs or small trees, 1-10 m tall. Many-branched, armed with opposite axillary stout thorns. Leaves opposite or crowded on very short lateral branches, leaf blade papery or subleathery, obovate or oblong-obovate, lateral veins 4-7 on each side of midvein. Flowers solitary or 2-3 clustered at apex of lateral short branches or tufts of leaves, calyx-tube 5-lobes. Berry globose, with grooves, 2-4 cm in diam. Fl. Mar.-Jun., fr. May-Jan. of next year.

花 Flower
摄影：吴林芳 Photo by: Wu Linfang

叶 Leaf
摄影：吴林芳 Photo by: Wu Linfang

果枝 Fruiting branch
摄影：吴林芳 Photo by: Wu Linfang

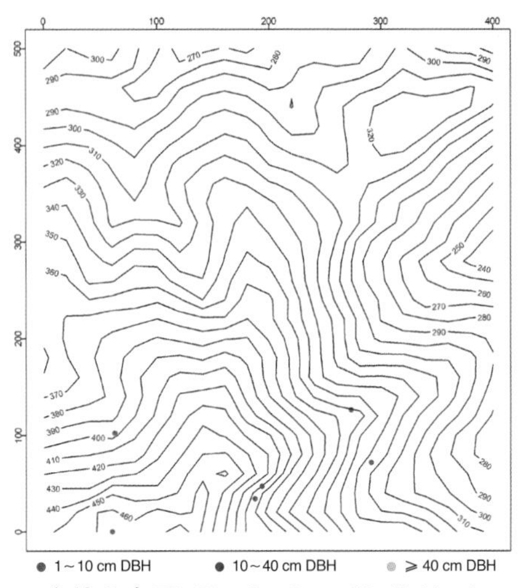

个体分布图 Distribution of individuals

径级分布表 DBH class

胸径等级 (Diameter class) (cm)	个体数 (No. of individuals in the plot)	比例 (Proportion) (%)
1～2	2	25.00
2～5	3	37.50
5～10	2	25.00
10～20	1	12.50
20～30	0	0.00
30～60	0	0.00
≥60	0	0.00

178 狗骨柴 gǒugǔchái | Common Tricalysia

Diplospora dubia (Lindl.) Masamune
茜草科 | Rubiaceae

代码（SpCode）= DIPDUB
个体数（Individual number/20 hm^2）= 317
最大胸径（Max DBH）= 16.9 cm
重要值排序（Importance value rank）= 11

常绿灌木或小乔木，高1～12m。无刺。小枝及叶无毛。叶对生，革质，卵状长圆形、长圆形、椭圆形或披针形，全缘。花多数簇生叶腋呈聚伞状，4基数，花萼筒顶端4齿，子房2室。浆果近球形，径4～6mm。花期1～5月，果期5月至翌年2月。

Evergreen shrubs or small trees, 1-12 m tall. Unarmed. Branches and leaves glabrous. Leaves opposite, leathery, ovate-oblong, oblong, elliptic or lanceolate, margin entire. Fllowers clustered in short dense axillary cymes, 4-merous, calyx-tube 4-toothed, ovary 2-celled. Berry subglobose, 4-6 mm in diam.. Fl. Jan.-May, fr. May-Feb. of next year.

果 Fruit
摄影：吴林芳 Photo by: Wu Linfang

花 Flower
摄影：吴林芳 Photo by: Wu Linfang

果枝 Fruiting branch
摄影：吴林芳 Photo by: Wu Linfang

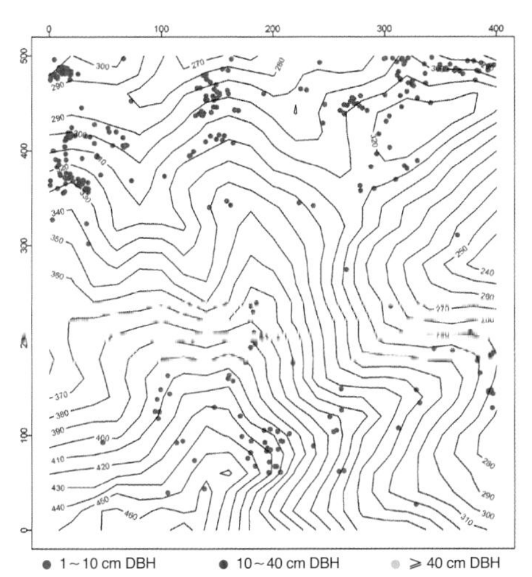

个体分布图 Distribution of individuals

径级分布表 DBH class

胸径等级 (Diameter class) (cm)	个体数 (No. of individuals in the plot)	比例 (Proportion) (%)
1～2	148	46.69
2～5	125	39.43
5～10	41	12.93
10～20	3	0.95
20～30	0	0.00
30～60	0	0.00
≥60	0	0.00

179 栀子

zhīzǐ | Cape Jasmine

Gardenia jasminoides Ellis
茜草科 | Rubiaceae

代码（SpCode）= GARJAS
个体数（Individual number/20 hm^2）= 32
最大胸径（Max DBH）= 2.9 cm
重要值排序（Importance value rank）= 85

常绿灌木，高1～3m。无刺。全株无毛或幼枝被粉。叶对生或少数3片轮生，近革质；椭圆形、长椭圆形或长披针形。花顶生，单个，花萼筒顶具5～6裂片，子房1室。果椭圆体，被宿存花萼筒包围，具多棱。花期4～8月，果期5～12月。

Evergreen shrubs, 1-3 m tall. Unarmed. Globrous throughout or young shoots pulverulent. Leaves opposite, or rarely in whorls of 3, subleathery, leaf blade elliptic, oblong-elliptic, or oblong-lanceolate. Flowers terminal, solitary, calyx-tube 5-6 sharp ribs, ovary 1-locular. Fruit ellipsoidal, crowned by persistent calyx-lobes with many ribs. Fl. Apr.-Aug., fr. May-Dec..

果 Fruit
摄影：吴林芳 Photo by: Wu Linfang

花 Flower
摄影：吴林芳 Photo by: Wu Linfang

花枝 Flowering branch
摄影：吴林芳 Photo by: Wu Linfang

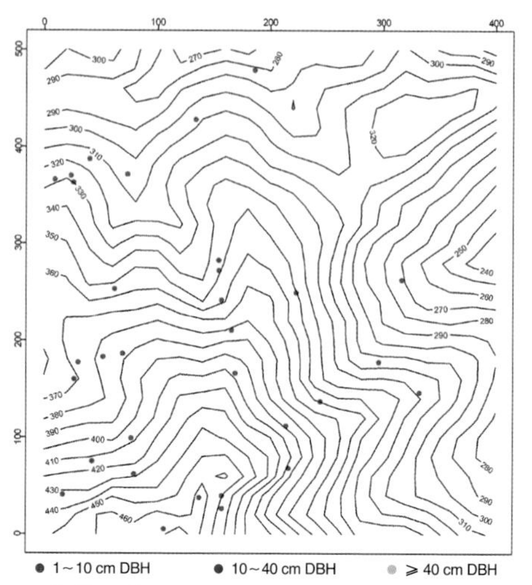

个体分布图 Distribution of individuals

径级分布表 DBH class

胸径等级 (Diameter class) (cm)	个体数 (No. of individuals in the plot)	比例 (Proportion) (%)
1～2	24	75
2～5	8	25
5～10	0	0.00
10～20	0	0.00
20～30	0	0.00
30～60	0	0.00
≥60	0	0.00

180 龙船花

lóngchuánhuā | Chinese Ixora

Ixora chinensis Lam.
茜草科 | Rubiaceae

代码（SpCode）= IXOCHI
个体数（Individual number/20 hm^2）= 1
最大胸径（Max DBH）= 1.0 cm
重要值排序（Importance value rank）= 195

常绿灌木，全株无毛。叶对生，近革质；披针形、长披针形，或长倒披针形。聚伞花序顶生，三分；花4基数；花萼筒顶具4裂片，子房2室，每室1种子；花冠鲜红色至橙色。核果球形，暗红色。花期7～11月。

Evergreen shrubs, glabrous throughout. Leaves opposite, subleathery, leaf blade lanceolate, oblong-lanceolate or oblong-oblanceolate. Cymes terminal, trichotomous, flower 4-merous, calyx-tube with 4-lobes, ovary 2-locular, with 1 ovule in each locule, corolla scarlet-orange. Drupe globular, reddish black when mature. Fl. Jul.-Nov..

叶 Leaf
摄影：吴林芳 Photo by: Wu Linfang

果序 Infructescence
摄影：吴林芳 Photo by: Wu Linfang

花 Flower
摄影：吴林芳 Photo by: Wu Linfang

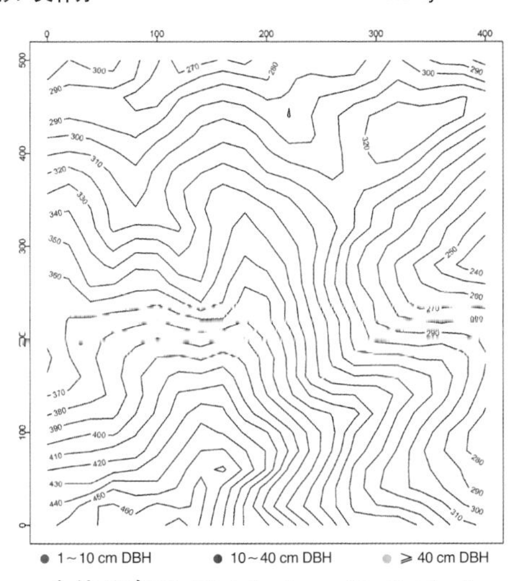

个体分布图 Distribution of individuals

径级分布表 DBH class

胸径等级 (Diameter class) (cm)	个体数 (No. of individuals in the plot)	比例 (Proportion) (%)
1～2	1	100.00
2～5	0	0.00
5～10	0	0.00
10～20	0	0.00
20～30	0	0.00
30～60	0	0.00
≥60	0	0.00

181 斜基粗叶木

xiéjīcūyèmù | Curtis lasianthus

Lasianthus attenuatus Jack.
茜草科 | Rubiaceae

代码（SpCode）= LASATT
个体数（Individual number/20 hm^2）= 3
最大胸径（Max DBH）= 2.0 cm
重要值排序（Importance value rank）= 168

常绿灌木。小枝密被刚毛。叶对生，革质，叶基明显偏斜，长披针形或长圆形，侧脉6～8对。花几朵簇生叶腋；花萼裂片近等长于萼管；子房5室。核果近球形或卵形，蓝至蓝黑色。花期4月，果期8～9月。

Evergreen shrubs. Branchlets densely hirsute. Leaves opposite, leathery, leaf base markedly oblique, leaf blade oblong-lanceolate or oblong, lateral veins 6-8 on each side of midvein. Flowers fewer cluster in axillary sessile clusters, calyx lobes as long as, ovary 5-locular. Drupes subglobose, blue to dark blue. Fl. Apr., fr. Aug.-Sep..

叶背 Leaf abaxial surface
摄影：吴林芳 Photo by: Wu Linfang

花 Flower
摄影：吴林芳 Photo by: Wu Linfang

叶 Leaf
摄影：吴林芳 Photo by: Wu Linfang

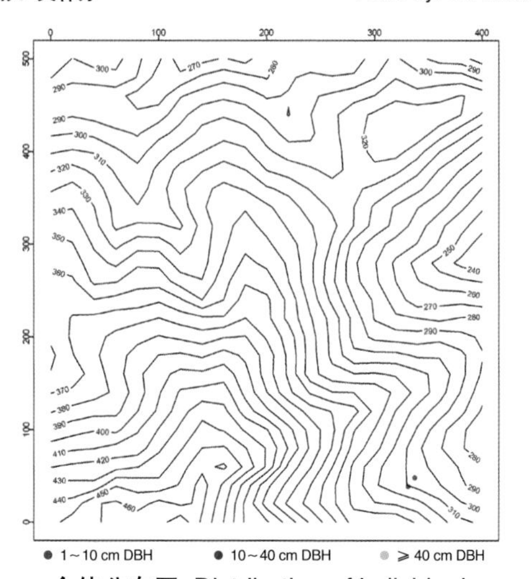

个体分布图 Distribution of individuals

径级分布表 DBH class

胸径等级 (Diameter class) (cm)	个体数 (No. of individuals in the plot)	比例 (Proportion) (%)
1～2	3	100.00
2～5	0	0.00
5～10	0	0.00
10～20	0	0.00
20～30	0	0.00
30～60	0	0.00
≥60	0	0.00

182 粗叶木

cūyèmù | Chinese Lasianthus

Lasianthus chinensis Benth.

茜草科 | Rubiaceae

代码（SpCode）= LASCHI

个体数（Individual number/20 hm^2）= 27

最大胸径（Max DBH）= 2.6 cm

重要值排序（Importance value rank）= 72

常绿灌木，高2～4m。嫩枝扁而被毛，老枝圆而无毛。叶对生，薄革质；长圆形、椭圆形或长披针形，侧脉9～14对。花5～6朵簇生叶腋，花萼裂片短于萼管，子房6室，每室1胚珠。核果球形，蓝紫色。花期4～12月，果期5～12月。

Evergreen shrubs, 2-4 m tall. Branches flattened and shortly tomentose when young, terete and glabrous in age. Leaves opposite, thinly leathery, leaf blade oblong, elliptic, or oblong-lanceolate, lateral veins 9-14 on each side of midvein. Flowers 5-6 in axillary sessile clusters, calyx lobes shorter than tube, ovary 6-locular, with 1 ovule in each locule. Drupes globular, blue or purplish. Fl. Apr.-Dec. fr. May-Dec..

叶 Leaf
摄影：吴林芳 Photo by: Wu Linfang

果 Fruit
摄影：吴林芳 Photo by: Wu Linfang

花 Flower
摄影：吴林芳 Photo by: Wu Linfang

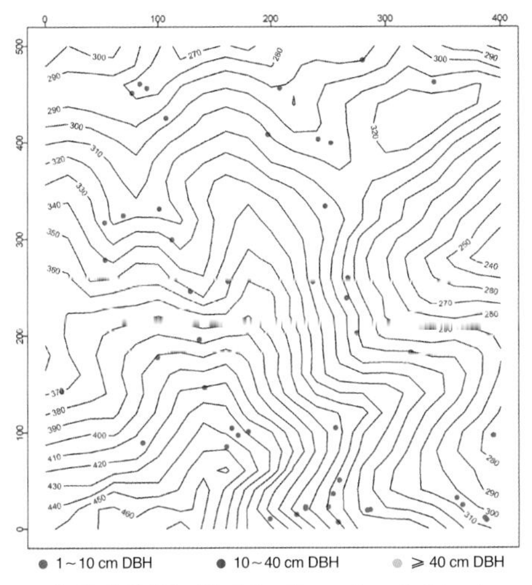

个体分布图 Distribution of individuals

径级分布表 DBH class

胸径等级 (Diameter class) (cm)	个体数 (No. of individuals in the plot)	比例 (Proportion) (%)
1～2	25	92.59
2～5	2	7.41
5～10	0	0.00
10～20	0	0.00
20～30	0	0.00
30～60	0	0.00
≥60	0	0.00

183 乌檀

wūtán | Medicinal Fatheadtree

Nauclea officinalis (Pierre ex Pit.) Merr. & Chun
茜草科 | Rubiaceae

代码（SpCode）= NAUOFF
个体数（Individual number/20 hm²）= 9
最大胸径（Max DBH）= 43.8 cm
重要值排序（Importance value rank）= 102

常绿乔木，高4～12m。小枝纤细，光滑。顶芽压扁，倒卵形。叶对生，纸质，椭圆形，少数倒卵形。头状花序单个顶生，子房2室，每室多粒种子。果头状，肉质，黄褐色，含许多小核果。花期夏季，果期4～12月。

Evergreen trees, 4-12 m tall. Branchlets slender, smooth. Terminal bud strongly compressed, obovoid. Leaves opposite, papery, leaf blade elliptic, rarely obovate. Flowering head solitary, terminal, ovary 2-locular, with numerous ovules in each locule. Fruiting head globular, fleshy, yellowish brown, comprised of numerous drupelets. Fl. summer, fr. Apr.-Dec..

野外识别特征：

1. 叶较大，长7～11(～15)cm，宽3.5～7(～10)cm；
2. 叶脉在叶面明显凹下，侧脉5～12对；
3. 叶对生，纸质。

Key notes for identification:

1. Leaf blade relatively big, 7-11 (-15) cm × 3.5-7(-10) cm.
2. Veins usually conspicuous impressed adaxially, secondary veins 5-12 pairs.
3. Leaves opposite, papery.

枝叶 Branch and leaves
摄影：吴林芳 Photo by: Wu Linfang

叶 Leaf
摄影：吴林芳 Photo by: Wu Linfang

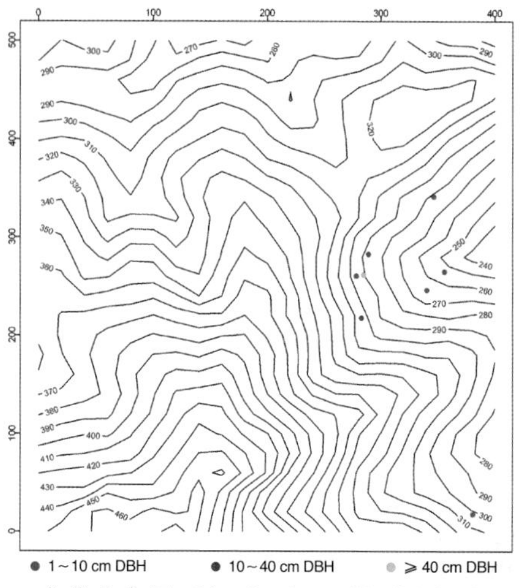

个体分布图 Distribution of individuals

径级分布表 DBH class

胸径等级 (Diameter class) (cm)	个体数 (No. of individuals in the plot)	比例 (Proportion) (%)
1～2	1	11.11
2～5	4	44.44
5～10	0	0.00
10～20	1	11.11
20～30	1	11.11
30～60	2	22.22
≥60	0	0.00

184 香港大沙叶 xiānggǎngdàshāyè | Hongkong Pavetta

Pavetta hongkongensis Brem.
茜草科 | Rubiaceae

代码（SpCode）= PAVHON
个体数（Individual number/20 hm^2）= 12
最大胸径（Max DBH）= 5.6 cm
重要值排序（Importance value rank）= 99

常绿灌木。小枝无毛。叶对生，薄纸质，常有菌瘤，长圆形，椭圆状倒卵形。聚伞花序3歧，顶生。花4基数，花萼裂片短于萼管，子房2室，每室1种子。浆果球形，径约6mm。花期3～10月，果期6～12月。

Evergreen shrubs. Branchlets glabrous. Leaves opposite, thinly papery, usually with batonoma, leaf blade oblong, elliptic-obovate. Cymes trichotomous, sessile above the last leaves. Flowers 4-merous, calyx lobes shorter than tube, ovary 2-locular, with 1 ovule in each locule. Berry globular, ca. 6 mm in diam.. Fl. Mar.-Oct., fr. Jun.-Dec..

花 Flower
摄影：吴林芳 Photo by: Wu Linfang

果枝 Fruiting branch
摄影：吴林芳 Photo by: Wu Linfang

叶 Leaf
摄影：吴林芳 Photo by: Wu Linfang

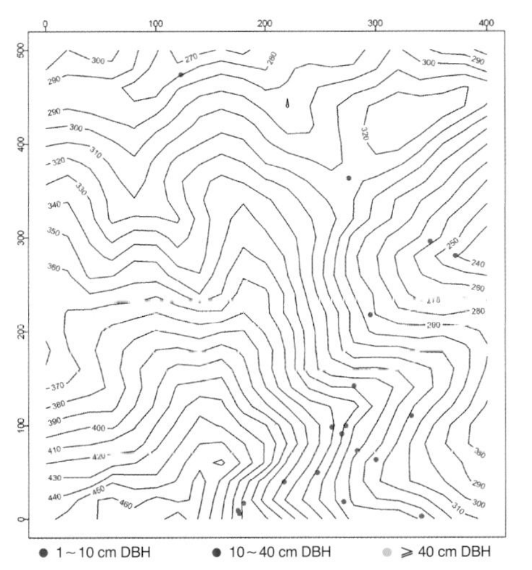

个体分布图 Distribution of individuals

径级分布表 DBH class

胸径等级 (Diameter class) (cm)	个体数 (No. of individuals in the plot)	比例 (Proportion) (%)
1～2	2	16.67
2～5	7	58.33
5～10	3	25.00
10～20	0	0.00
20～30	0	0.00
30～60	0	0.00
≥60	0	0.00

185 九节

jiǔjié | Red Psychotria

Psychotria asiatica Linn.

茜草科 | Rubiaceae

代码（SpCode）= PSYASI

个体数（Individual number/20 hm^2）= 853

最大胸径（Max DBH）= 9.3 cm

重要值排序（Importance value rank）= 26

常绿灌木或小乔木，高0.5～5m。叶对生，纸质或近革质，长圆形、椭圆状长圆形或长倒披针形，全缘，无毛。聚伞花序顶生，常呈伞状或圆锥状，一般三歧，花萼管极短，子房2室，每室1种子。核果球形或大体椭圆体，具纵纹，熟时红色。花期3～9月，果期7月至翌年2月。

Evergreen shrubs or small trees, 0.5-5 m tall. Leaves opposite, papery or subleathery, leaf blade oblong, elliptic-oblong or oblong-oblanceolate, margin entire, glabrous. Cymes corymbiform or paniculate, terminal, usually trichotomous, calyx-tube very short, ovary 2-locular, with 1 ovule in each locule. Drupe globose or broadly ellipsoidal, with longitudinal ribs, red when mature. Fl. Mar.-Sep., fr. Jul.-Feb. of next year.

花 Flower
摄影：吴林芳 Photo by: Wu Linfang

果枝 Fruiting branch
摄影：吴林芳 Photo by: Wu Linfang

叶 Leaf
摄影：吴林芳 Photo by: Wu Linfang

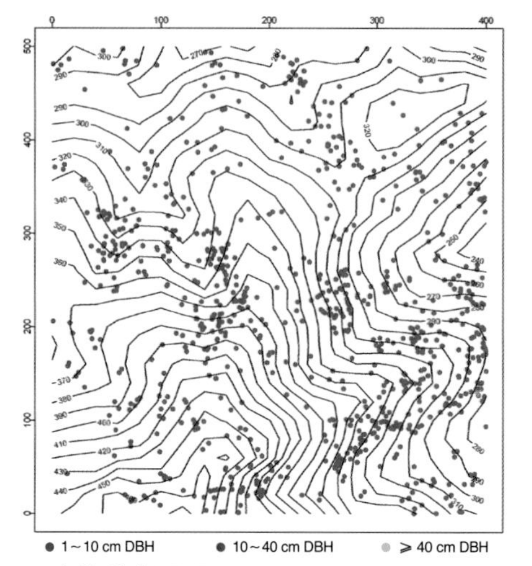

个体分布图 Distribution of individuals

径级分布表 DBH class

胸径等级 (Diameter class) (cm)	个体数 (No. of individuals in the plot)	比例 (Proportion) (%)
1～2	545	63.89
2～5	301	35.29
5～10	7	0.82
10～20	0	0.00
20～30	0	0.00
30～60	0	0.00
≥60	0	0.00

186 密毛乌口树（白花苦灯笼） mìmáowūkǒushù | Whiteflower Tarenna

Tarenna mollissima (Hook. et Arn.) Robins.
茜草科 | Rubiaceae

代码（SpCode）= TARMOL
个体数（Individual number/20 hm²）= 26
最大胸径（Max DBH）= 2.9 cm
重要值排序（Importance value rank）= 76

常绿灌木或小乔木，高1~6m，各部分密被柔毛。叶对生，纸质，披针形、长披针形或卵状椭圆形，全缘。聚伞花序顶生，3歧，花萼裂片与萼管等长，子房2室，每室1种子。浆果球形，黑色，被毛。花期5~7月，果期7月至翌年2月。

Evergreen shrubs or small trees, 1-6 m tall. Softly and densely pubescent in every part. Leaves opposite, papery, leaf blade lanceolate, oblong-lanceolate or ovate-elliptic, margin entrie. Cymes corymbiform, trichotomous, terminal, calyx lobes as length as tube, ovary 2-locular, with 1 ovules in each locule. Berry globular, black, pubescent. Fl. May-Jul., fr. Jul.-Fed. of next year.

幼苗 Seedling
摄影：吴林芳 Photo by: Wu Linfang

花 Flower
摄影：吴林芳 Photo by: Wu Linfang

果 Fruit
摄影：吴林芳 Photo by: Wu Linfang

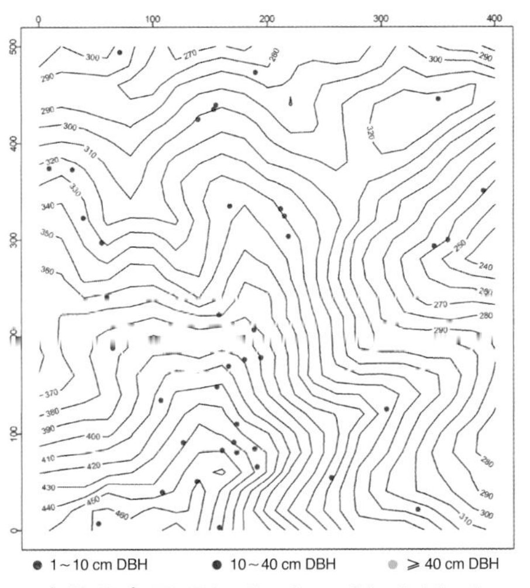

个体分布图 Distribution of individuals

径级分布表 DBH class

胸径等级 (Diameter class) (cm)	个体数 (No. of individuals in the plot)	比例 (Proportion) (%)
1~2	23	88.46
2~5	3	11.54
5~10	0	0.00
10~20	0	0.00
20~30	0	0.00
30~60	0	0.00
≥60	0	0.00

187 坚荚树（常绿荚蒾） jiānjiáshù | Evergreen Viburnum

Viburnum sempervirens K. Koch
忍冬科 | Caprifoliaceae

代码（SpCode）= VIBSEM
个体数（Individual number/20 hm^2）= 1
最大胸径（Max DBH）= 1.2 cm
重要值排序（Importance value rank）= 169

常绿灌木，高2～4m。小枝紫色或灰褐色。叶对生，革质，椭圆形至椭圆状卵形，全缘或尾端略具齿，侧脉每边3～4条，有时5条。聚伞花序排成复伞形花序式，顶生。核果红色，近球形或卵球形。花期4～5月，果期8～11月。

Evergreen shrubs, 2-4 m tall. Branchlets purple or grey-brown. Leaves opposite, leathery, leaf blade elliptic to elliptic-ovate, margin entire or slightly toothed towards end, lateral veins 3-4 (-5) on each side of midvein. Cymes arranged into compound umbel, terminal. Drupes red, subglobular or ovoid. Fl. Apr.-May, fr. Aug.-Nov..

花 Flower
摄影：卓书斌 Photo by: Zuo Shubin

果 Fruit
摄影：吴林芳 Photo by: Wu Linfang

叶 Leaf
摄影：吴林芳 Photo by: Wu Linfang

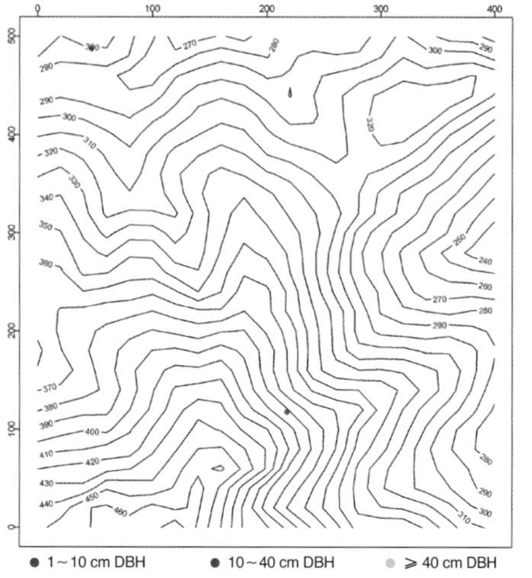

个体分布图 Distribution of individuals

径级分布表 DBH class

胸径等级 (Diameter class) (cm)	个体数 (No. of individuals in the plot)	比例 (Proportion) (%)
1～2	1	100.00
2～5	0	0.00
5～10	0	0.00
10～20	0	0.00
20～30	0	0.00
30～60	0	0.00
≥60	0	0.00

188 破布木 pòbùmù | Dichotomous Cordia

Cordia dichotoma G. Forst.
紫草科 | Boraginaceae

代码（SpCode）= CORDIC
个体数（Individual number/20 hm²）= 1
最大胸径（Max DBH）= 1.7 cm
重要值排序（Importance value rank）= 185

落叶乔木，高3~8m。小枝幼时被疏毛，后无毛。叶互生，薄革质，卵形至阔卵形或阔椭圆形，略被毛或无毛。伞房状聚伞花序生于具叶的侧枝顶端，花柱两次2裂。核果黄色或淡红色，近球形，种子1。花期4~6月，果期6~9月。

Deciduous trees, 3-8 m tall. Branches sparsely pilose when young, glabrescent. Leaves alternate, thinly leathery, ovate to broadly ovate ot elliptic, sparsely pubescent or glabrous. Inflorescences terminating leafy lateral branches, dichotomously branched into corymbose cymes, style twice 2-cleft. Drupes yellow or reddish, subglobose, 1-seed. Fl. Apr.-Jun., fr. Jun.-Sep..

果枝 Fruiting branch
摄影：吴林芳 Photo by: Wu Linfang

枝叶 Branch and leaves
摄影：吴林芳 Photo by: Wu Linfang

叶背 Leaf abaxial surface
摄影：吴林芳 Photo by: Wu Linfang

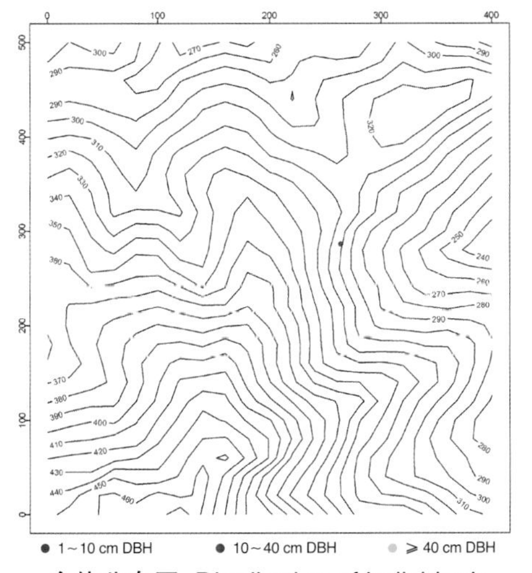

● 1~10 cm DBH ● 10~40 cm DBH ● ≥ 40 cm DBH
个体分布图 Distribution of individuals

径级分布表 DBH class

胸径等级 (Diameter class) (cm)	个体数 (No. of individuals in the plot)	比例 (Proportion) (%)
1~2	1	100.00
2~5	0	0.00
5~10	0	0.00
10~20	0	0.00
20~30	0	0.00
30~60	0	0.00
≥60	0	0.00

189 长花厚壳树

chánghuāhòukéshù | Longflower Ehretia

Ehretia longiflora Champ. ex Benth.
紫草科 | Boraginaceae

代码（SpCode）= EHRLON
个体数（Individual number/20 hm^2）= 1
最大胸径（Max DBH）= 1.7 cm
重要值排序（Importance value rank）= 186

落叶乔木，高5～15m。树皮暗灰色或暗褐色，片状剥落。小枝无毛。叶纸质，椭圆形至长圆形或长倒披针形，全缘，无毛。伞状聚伞花序顶生，花柱2裂，柱头2。核果淡黄至红色，无毛，内果皮有棱，分成4个具1种子的小坚果。花果期1～9月。

Deciduous trees, 5-15m tall. Bark dark grey to dark brown, scaly. Branchlets glabrous. Leaf blade papery, elliptic to oblong or oblong-oblanceolate, margin entire, glabrous. Cymes corymbose, terminating later branches, style 2-cleft, stigmas 2. Drupes pale yellow to red, endocarp ribbed, divided into four 1-seeded pyrenes. Fl. and fr. Jan.-Sep..

果 Fruit
摄影：吴林芳 Photo by: Wu Linfang

花 Flower
摄影：吴林芳 Photo by: Wu Linfang

叶 Leaf
摄影：吴林芳 Photo by: Wu Linfang

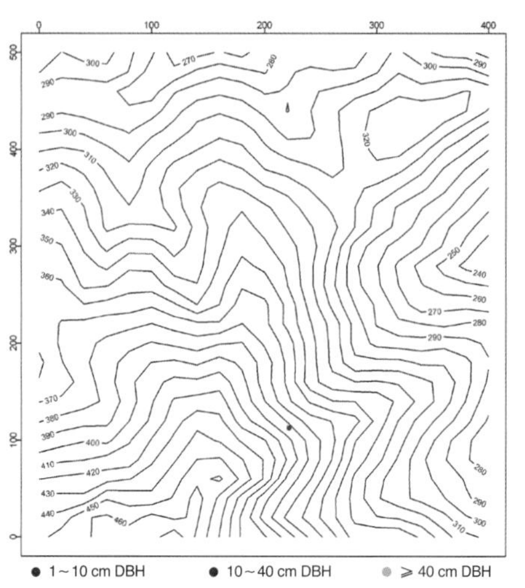

个体分布图 Distribution of individuals

径级分布表 DBH class

胸径等级 (Diameter class) (cm)	个体数 (No. of individuals in the plot)	比例 (Proportion) (%)
1～2	1	100.00
2～5	0	0.00
5～10	0	0.00
10～20	0	0.00
20～30	0	0.00
30～60	0	0.00
≥60	0	0.00

190 大青 dàqīng | Mayflower Glorybery

Clerodendrum cyrtophyllum Turcz.

马鞭草科 | Verbenaceae

代码（SpCode）= CLECYR

个体数（Individual number/20 hm²）= 1

最大胸径（Max DBH）= 1.6 cm

重要值排序（Importance value rank）= 187

常绿灌木或小乔木，高1～10m。小枝黄褐色，被毛。叶对生，叶纸质，椭圆形、卵状椭圆形、长圆形或长圆状披针形，全缘。伞状聚伞花序顶生或近顶生，花小，芳香；花萼杯状，5裂，花冠白色，冠管明显长于花萼。核果蓝紫色，卵球形至球形，花果期6月至翌年2月。

Evergreen shrubs or small trees, 1-10 m tall. Branchlets yellowish brown, pubescent. Leaves simple opposite, leaf blade papery, elliptic, ovate-elliptic, oblong or oblong-lanceolate, margin entire. Cymes corymbose, terminal or subterminal. Flowers small, fragrant. Calyx copular, 5-lobes, corolla white, tube longer than calyx. Drupes blue-purple, obovoid to globose. Fl. & fr. Jun.-Feb. of next year.

叶背 Leaf abaxial surface

摄影：吴林芳 Photo by: Wu Linfang

果 Fruit

摄影：吴林芳 Photo by: Wu Linfang

花 Flower

摄影：吴林芳 Photo by: Wu Linfang

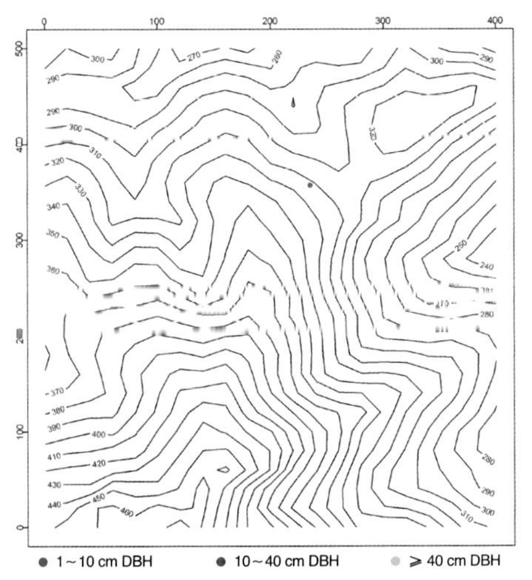

个体分布图 Distribution of individuals

径级分布表 DBH class

胸径等级 (Diameter class) (cm)	个体数 (No. of individuals in the plot)	比例 (Proportion) (%)
1～2	0	0.00
2～5	1	100.00
5～10	0	0.00
10～20	0	0.00
20～30	0	0.00
30～60	0	0.00
≥60	0	0.00

191 鬼灯笼（白花灯笼） guǐdēnglóng | Gloryberry

Clerodendrum fortunatum L.
马鞭草科 | Verbenaceae

枝叶 Branch and leaves
摄影：吴林芳 Photo by: Wu Linfang

代码（SpCode）= CLEFOR
个体数（Individual number/20 hm^2）= 2
最大胸径（Max DBH）= 2.7 cm
重要值排序（Importance value rank）= 166

半落叶灌木。嫩枝密被黄褐色短柔毛。单叶对生，纸质，长椭圆形、倒卵状披针形，叶背密布黄色小腺点。聚伞花序腋生，比叶短，花萼紫红色，5棱，冠管略长于花萼。核果深蓝色，近球形，藏于宿萼内。花果期6～11月。

Semideciduous shrubs. Twigs densely yellowish brown pubescent. Leaves simple opposite. Leaf blade papery, long elliptic, obovate-lanceolate, abaxially densely small yellow glandular. Cymes axillary, shorter than leave, calyx reddish purple, 5-veined, corolla slightly longer than calyx. Drupes dark blue-green, subglobose, enxlosed by persistent calyx. Fl. & fr. Jun.-Nov..

花 Flower
摄影：吴林芳 Photo by: Wu Linfang

果 Fruit
摄影：吴林芳 Photo by: Wu Linfang

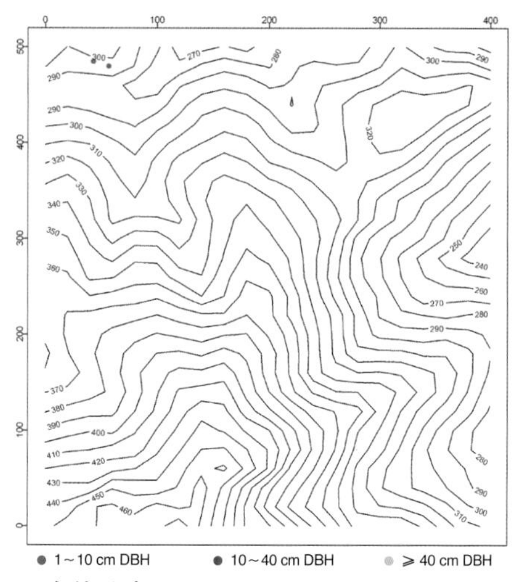

个体分布图 Distribution of individuals

径级分布表 DBH class

胸径等级 (Diameter class) (cm)	个体数 (No. of individuals in the plot)	比例 (Proportion) (%)
1～2	1	50.00
2～5	1	50.00
5～10	0	0.00
10～20	0	0.00
20～30	0	0.00
30～60	0	0.00
≥60	0	0.00

192 赪桐 chēngtóng | Pagoda Flower

Clerodendrum japonicum (Thunb.) Sweet
马鞭草科 | Verbenaceae

代码（SpCode）= CLEJAP
个体数（Individual number/20 hm^2）= 2
最大胸径（Max DBH）= 2.5 cm
重要值排序（Importance value rank）= 167

半常绿灌木。小枝四方形，被毛或近无毛。单叶对生，纸质，近心形，叶背密被盾状腺点和黄褐色柔毛。聚伞圆锥花序顶生，花萼红色，深5裂，花冠红色，明显长于花萼，雄蕊及花柱是花冠管的3倍或多。核果近球形，蓝黑色，花果期5～11月。

Semievergreen shrubs. Branchlets tetragonal, pubscent to subglabrous. Leaves simple opposite, leaf blade papery, nearly heart-shaped, abaxially densely peltate glands and yellowish brown pubscent. Thyrses terminal, calyx red, deeply 5-lobed, corolla red, clearly longer than calyx, stamens and style 3 times as long as corolla-tube. Drupes subglobose, bluish black. Fl. and fr. May-Nov..

花 Flower
摄影：吴林芳 Photo by: Wu Linfang

叶 Leaf
摄影：吴林芳 Photo by: Wu Linfang

果 Fruit
摄影：吴林芳 Photo by: Wu Linfang

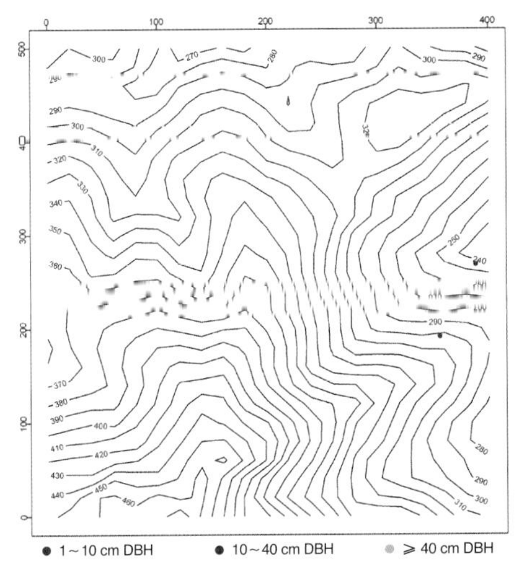

个体分布图 Distribution of individuals

径级分布表 DBH class

胸径等级 (Diameter class) (cm)	个体数 (No. of individuals in the plot)	比例 (Proportion) (%)
1~2	1	50
2~5	1	50
5~10	0	0
10~20	0	0
20~30	0	0
30~60	0	0
≥60	0	0

193 山牡荆

shānmǔjīng | Wild Vitex

Vitex quinata (Lour.) Will
马鞭草科 | Verbenaceae

代码（SpCode）= VITQUI
个体数（Individual number/20 hm^2）= 12
最大胸径（Max DBH）= 21.6 cm
重要值排序（Importance value rank）= 101

常绿乔木，高4～20m。树皮褐色。小枝四棱，被柔毛和腺点。叶掌状复叶，3小叶，少数5小叶，厚纸质，倒卵状披针形至椭圆状披针形，全缘，叶背被黄色腺点。圆锥花序顶生，密被黄褐色柔毛。核果卵球形或球形，黑色。花期5～7月，果期9～11月。

Evergreen trees, 4-20 m tall. Bark brown. Branchlets tetrabonal, pubescent ang glandular. Leaves 3(-5)-foliolate, leaflets thickly papery, obovate-lanceolate to elliptic-lanceolate, margina entire, abaxially yellow glandular. Panicles terminal, densely yellowish brown pubescent. Drupes obovoid or globose, black. Fl. May-Jul., fr. Sep.-Nov..

果 Fruit
摄影：吴林芳 Photo by: Wu Linfang

花 Flower
摄影：吴林芳 Photo by: Wu Linfang

枝叶 Branch and leaves
摄影：吴林芳 Photo by: Wu Linfang

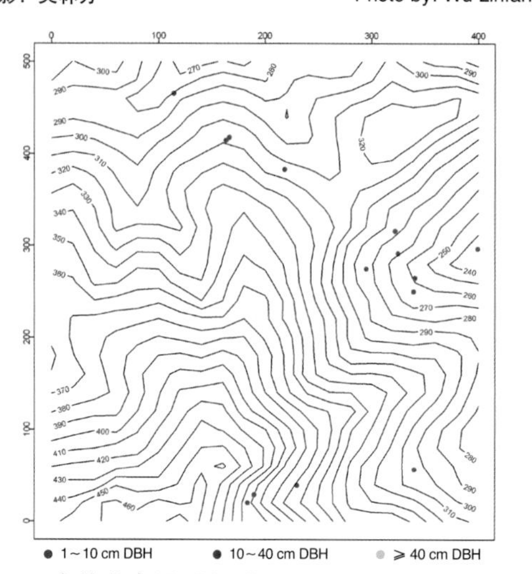

个体分布图 Distribution of individuals

径级分布表 DBH class

胸径等级 (Diameter class) (cm)	个体数 (No. of individuals in the plot)	比例 (Proportion) (%)
1～2	0	0.00
2～5	1	8.33
5～10	5	41.67
10～20	4	33.33
20～30	2	16.67
30～60	0	0.00
≥60	0	0.00

194 鱼尾葵 yúwěikuí | Fishtail Palm

Caryota maxima Blume
棕榈科 | Arecaceae

代码（SpCode）= CARMAX
个体数（Individual number/20 hm^2）= 38
最大胸径（Max DBH）= 30.5 cm
重要值排序（Importance value rank）= 66

常绿乔木，高达20m。干光滑，有环状叶痕，单生。叶二回羽状全裂，裂片顶端不规则齿缺，侧面裂片菱形而似鱼尾。佛焰苞和花序无鳞秕；肉穗花序长约3m；分枝悬垂。果球形，径约2cm，熟时淡红或紫红色。花期7月，果期1～2年后熟。

Evergreen trees up to 20 m tall. Bark glossy, with orbicular cicatricle, single. Leaves bipinnate, complete cleavage, apex of sliver irregularly nicks, lateral slivers rhombus like fishtail. Spathe and inflorescence with no peltate scale, spadix ca. 3 m, ramose, overhanged. Fruits globose, ca. 2 cm in diam., reddish or red-purple. Fl. Jul., fr. 1-2 years later.

植株 whole plant
摄影：吴林芳 Photo by: Wu Linfang

叶 Leaves
摄影：吴林芳 Photo by: Wu Linfang

幼苗 Seedling
摄影：吴林芳 Photo by: Wu Linfang

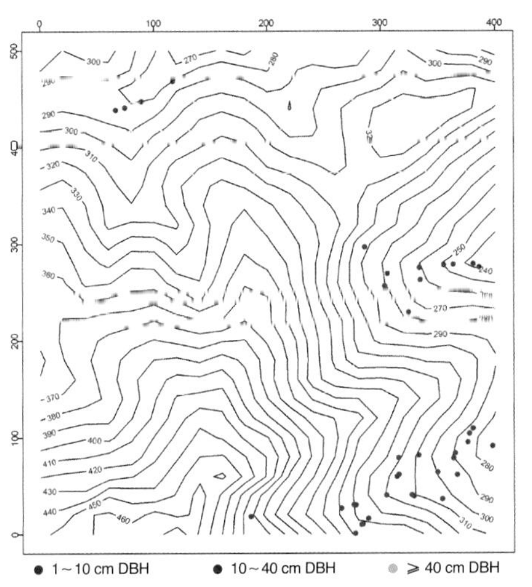

个体分布图 Distribution of individuals

径级分布表 DBH class

胸径等级 (Diameter class) (cm)	个体数 (No. of individuals in the plot)	比例 (Proportion) (%)
1～2	0	0.00
2～5	0	0.00
5～10	3	7.89
10～20	17	44.74
20～30	17	44.74
30～60	1	2.63
≥60	0	0.00

195 茶秆竹

chágănzhú | Tonkin Cane

Pseudosasa amabilis (Mc Clure) P.C.Keng ex S.L.Chen
禾本科 | Gramineae

代码（SpCode）= PSEAMA
个体数（Individual number/20 hm^2）= 19
最大胸径（Max DBH）= 4.4 cm
重要值排序（Importance value rank）= 133

常绿小乔木，秆高6～13m，直径2～6cm。节间圆柱形，秆环平坦或稍隆起。秆每节1～3分枝，箨鞘迟落，背面被棕色刺毛，箨耳缺；末级小枝具2～3片叶，叶长16～35cm，宽1.6～3.5cm。总状或圆锥花序具3～15小穗，小穗轴弯曲，雄蕊3。笋期3～5月。

Evergreen arborescent bamboos, culms 6-13 m, 2-6 cm in diam.. Internodes olive-green, terete, nodes weakly prominent. Branches (1-)3 per node, culm sheaths gradually deciduous, densely brown setose, auricles absent. Leaves 2 or 3 per ultimate branch, 16-35 cm × 1.6-3.5 cm. Inflorescence paniculate, lateral spikelets 3-15, rachilla sinuous, stamens 3. New shoots Mar. - May.

树干 Trunk
摄影：吴林芳 Photo by: Wu Linfang

枝叶 Branch and leaves
摄影：吴林芳 Photo by: Wu Linfang

叶 Leaf
摄影：吴林芳 Photo by: Wu Linfang

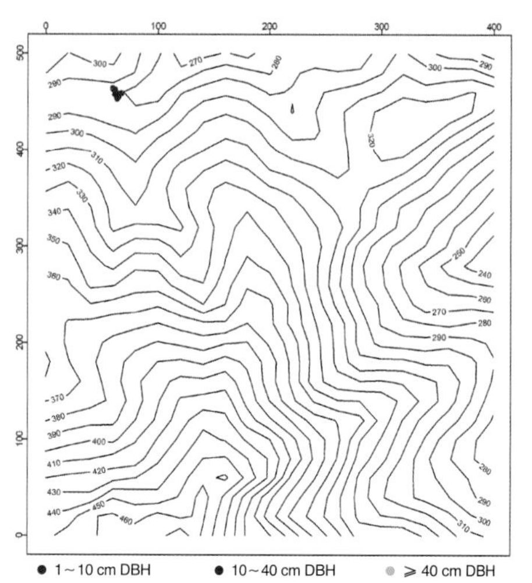

个体分布图 Distribution of individuals

径级分布表 DBH class

胸径等级 (Diameter class) (cm)	个体数 (No. of individuals in the plot)	比例 (Proportion) (%)
1～2	3	15.79
2～5	16	84.21
5～10	0	0.00
10～20	0	0.00
20～30	0	0.00
30～60	0	0.00
≥60	0	0.00

附录I 植物中文名索引
Appendix I Chinese Species Name Index

附录II 植物学名索引
Appendix II Scientific Species Name Index